Yuefei Zeng

Efficient Radar Forward Operator for Operational Data Assimilation within the COSMO-model

Wissenschaftliche Berichte des Instituts für Meteorologie und Klimaforschung des Karlsruher Instituts für Technologie (KIT)
Band 60

Herausgeber: Prof. Dr. Ch. Kottmeier

Institut für Meteorologie und Klimaforschung
am Karlsruher Institut für Technologie (KIT)
Kaiserstr. 12, 76128 Karlsruhe

Eine Übersicht über alle bisher in dieser Schriftenreihe erschienenen Bände finden Sie am Ende des Buchs.

Efficient Radar Forward Operator for Operational Data Assimilation within the COSMO-model

by
Yuefei Zeng

KIT Scientific Publishing

Dissertation, Karlsruher Institut für Technologie (KIT)
Fakultät für Physik, 2013
Referenten: Prof. Dr. Klaus D. Beheng, Prof. Dr. Christoph Kottmeier

Impressum

KIT Scientific Publishing

Karlsruher Institut für Technologie (KIT)
KIT Scientific Publishing
Straße am Forum 2
D-76131 Karlsruhe

KIT Scientific Publishing is a registered trademark of Karlsruhe Institute of Technology. Reprint using the book cover is not allowed.

www.ksp.kit.edu

Print on Demand 2014

ISSN 0179-5619
ISBN 978-3-7315-0128-2

Efficient Radar Forward Operator for Operational Data Assimilation within the COSMO-model

Zur Erlangung des akademischen Grades eines
DOKTORS DER NATURWISSENSCHAFTEN
von der Fakultät für Physik des
Karlsruher Instituts für Technologie (KIT)

genehmigte
DISSERTATION
von

Dipl.-Math. Yuefei Zeng
aus
Fujian, China

Tag der mündlichen Prüfung: 05.07.2013
Referent: Prof. Dr. Klaus D. Beheng
Korreferent: Prof. Dr. Christoph Kottmeier

To my loving parents

Abstract

The new weather radar network of the German Weather Service (DWD) will, after its complete update in 2014, comprise 17 dual-polarimetric C-Band Doppler radars evenly distributed throughout Germany for complete coverage. They provide unique 3-dimensional information about dynamical and microphysical characteristics of precipitating clouds in high spatial and temporal resolutions. Up to now, these data are not used in the operational COSMO-model of DWD except within the framework of the latent heat nudging and for a simple nudging method of Doppler velocity. Future applications are, however, planned to take better advantage of radar data within an upcoming new 4-Dimensional Local Ensemble Transform Kalman Filter (4D-LETKF) data assimilation system, which will be based on the operational convective-scale ensemble prediction system (EPS) COSMO-DE-EPS (grid spacing of 2.8 km, rapid update cycle, Central Europe domain). It is assumed that the assimilation of weather radar data is a promising means for improvements of short-term precipitation forecasts, especially in convective situations.

However, the observed quantities (reflectivity, Doppler velocity and polarization properties) are not directly comparable to the prognostic variables of the numerical model. In order to, on one hand, enable radar data assimilation in the framework of the above-mentioned 4D-LETKF-assimilation system and, on the other hand, to facilitate comparisons of numerical simulations with radar observations in the context of cloud microphysics verification, a comprehensive modular radar forward operator has been developed. This operator simulates the measurement process of radar observables from the prognostic cloud physical model variables and allows for direct comparison in terms of radar observables. The operator consists of several modules, each of which handles a special physical process (e.g., scattering, extinction, microwave propagation, etc.). Each of these modules offers different formulations associated with different complexity in simulations, which can be flexibly chosen according to user's needs. In order to assess the performance of the operator, a series of sensitive experiments have been conducted. The main goal here is to find an optimal configuration of the operator in the sense of balance between physical accuracy and computational expense. Examples of various possibilities which can be chosen depending on the situations are: 1) the radar beam can be considered to propagate as a simple ray or treated with the actual volume averaging characteristic; 2) beam bending can be either derived from a 4/3 earth radius concept or from the actual simulated vertical gradient of the refractive index of air; 3) radar reflectivity may be calculated from the full Mie-theory or from various (more efficient) approximations. 4) Attenuation effects may be taken into account or not. To meet operational demands, the operator should be compatible with supercomputer architectures. Moreover, the program code has to be as efficient as possible, which requires good parallelization and vectorization properties of the code.

The results of sensitivity experiments show that the operator is able to efficiently simulate reflectivity and Doppler velocity under consideration of effects like beam bending and broadening as well as attenuation.

After having developed the operator and integrated the processing of radar data into the 4D-LETKF software package provided by DWD, we have exemplarily performed first data assimilation experiments. For that we have investigated the convective event of 31 May 2011. The required COSMO-DE ensemble is driven by a test ensemble of the global model GME. The preliminary results are then presented.

Kurzfassung

Das neue Wetterradar-Netzwerk des Deutschen Wetterdiensts (DWD) wird nach seiner vollständigen Aktualisierung im Jahr 2014 17 dual-polarisierte C-Band Dopplerradare enthalten, die mit einer kompletten Abdeckung gleichmäßig in ganz Deutschland verteilt sind. Wetterradare liefern einzigartige 3-dimensionale Informationen über dynamische und mikrophysikalische Eigenschaften von Niederschlagswolken in hohen räumlichen und zeitlichen Auflösungen. Bisher werden diese Daten noch nicht im operationellen Wettervorhersagemodell COSMO des DWD verwendet außer bei einfachen Verfahren zum Nudging der latenten Wärme und der Dopplergeschwindigkeit. Jedoch ist eine Nutzung der Radardaten ist in der Zukunft geplant, nämlich indem Radardaten mit dem zukünftigen neuen Datenassimilationssystem 4-Dimensional Local Ensemble Transform Kalman Filter (4D-LETKF) assimiliert werden, das auf dem konvektion-erlaubenden Ensemble-Prognose-System (EPS) COSMO-DE-EPS (horizontaler Rasterabstand von 2,8 km, schneller Update-Zyklus, Zentraleuropa) basiert. Wir gehen davon aus, dass die Assimilation von Radardaten ein vielversprechendes Mittel zur Verbesserung der kurzfristigen Niederschlagsvorhersage ist, insbesondere in konvektiven Situationen.

Doch die Radardaten (Reflektivität, Dopplergeschwindigkeit und Polarisationsparameter) sind nicht direkt vergleichbar mit den prognostischen Variablen COSMO-Modells. Um einerseits Radardaten in das oben genannte 4D-LETKF-Assimilationssystem zu assimilieren, und um andererseits den Vergleich zwischen Daten der numerischen Simulationen und Radardaten im Rahmen der Verifikation der Wolkenmikrophysik zu erleichtern, wird ein umfassender modularer Radarvorwärtsoperator entwickelt. Der Operator simuliert den Messvorgang von Radardaten aus den prognostischen wolkenmikrophysikalischen Modellvariablen und ermöglicht den direkten Vergleich in Form von Radardaten. Der Operator besteht aus mehreren Modulen, die jeweils ein spezielles physikalisches Verfahren (z.B., Streuung, Extinktion, Ausstrahlungsausbreitung, usw.) beschreiben. Jedes dieser Module bietet verschiedene Formulierungen mit unterschiedlichen Simulationskomplexitäten, die nach Bedarf flexibel gewählt werden können. Um die Leistung des Operators zu überprüfen, werden mehrere Sensitivitätsexperimente wurden durchgeführt. Das Hauptziel ist dabei eine optimale Konfiguration im Sinne eines Gleichgewichts zwischen physikalische Genauigkeit und Rechenaufwand zu finden. So kann zum Beispiel: 1) Jede Radartrahlung kann der Einfachheit halber als eine einzelne Linie betrachtet werden oder als tatsächlichen Volumenmittelungseigenschaft behandelt werden. 2) Die Ausbreitungstrajektorie kann entweder aus einem einfachen Konzept vom 4/3 Erdradius abgeleitet werden oder aus Methoden, die auf der Berüsichtigung der vertikalen Gradienten des Brechungsindexes der Luft beruhen. 3) Reflektivität kann aus der Mie-Theorie oder aus den anderen effizienteren Näherungen berechnet werden. 4) Dämpfungseffekte werden berücksichtigt oder nicht. Um den Operator operationell betreiben zu können, sollte er kompatibel mit der Supercomputer-Architektur sein. Dabei ist die Effizienz ist ein wichtiges Entwurfkriterium, sodass auf eine gute Parallelisierung und Vektorisierung des Codes Wert gelegt wird.

Die Ergebnisse von Sensitivitätsexperimenten zeigen, dass der Operator in der Lage ist, Reflektivität und Dopplergeschwindigkeit unter Berücksichtigung der Effekte wie Strahlbeugung, Strahlausbreitung und Dämpfung effizient zu simulieren.

Nach der Entwicklung des Operators and Integration von Radardaten in das 4D-LETKF Software-Paket, sind wir soweit, die ersten Datenassimilationsexperimente durchzuführen. Dazu haben wir das konvektive Ereignis vom 31.05.2011untersucht. Das benötigte COSMO-DE Ensemble wird angetrieben von einem Testensemble des globalen Modells GME. Die vorläufigen Ergebnisse werden zum Schluss vorgestellt.

Contents

1. Introduction

1.1. Motivation and background

Reliable quantitative precipitation forecasts (QPFs) are regarded as one of the most challenging tasks in numerical weather prediction (NWP). Especially in case of heavy precipitation events associated with small-scale convection like thunderstorms, the current operational forecast models have their problems. In order to resolve clouds and to describe the microphysical processes more accurately at the convective-scale, the current NWP models are and will be in the near future replaced or complemented by a new generation of nonhydrostatic mesoscale models with a horizontal resolution of 1-3 km. For instance, Meteo-France is working on the model AROME (Application of Research to Operations at Mesoscale), while the american model WRF (Weather Research and Forecasting) is developed and operated by the National Center for Atmospheric Research (NCAR), the National Oceanic and Atmospheric Administration (NOAA), the Center for Analysis and Prediction of Storms (CAPS) and by many scientists from universities becomes very popular in the atmospheric and climate science community.

Seven European national meteorological services including German Weather Service (Deutscher Wetterdienst, abbr. DWD) cooperate in the framework of the Consortium for Small-scale Modeling (COSMO) for both operational and research applications. The COSMO-model is a limited area non-hydrostatic and fully compressible NWP model in advection form, initially known as "Local Model (LM)" (Steppeler et al., 2003). While hydrostatic models like the global model GME of DWD are restricted to a grid spacing larger than about 10 km, non-hydrostatic models could in principle be applied at an extremely high resolution, e.g., 100 m, to resolve convective events.

From a mathematical point of view, NWP constitutes an initial (and boundary) value problem, which means that in order to be able to predict what the weather looks like in a few hours or days, it is a prerequisite to know the present weather as precisely

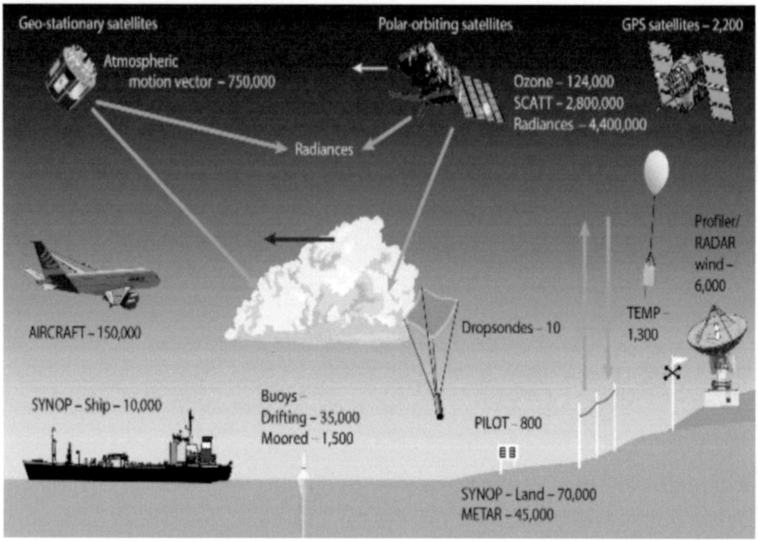

Fig. 1.1.: Meteorological observing system (Hagedorn, 2010)

as possible. Although a NWP model may provide the necessary framework for an explicit description of microphysical properties in the atmosphere due to its high resolution, improvements in forecasts can only be achieved if appropriate techniques for initialization of models based on observations in comparable spatial and temporal resolutions are also available. Data assimilation is presently the most popular one of these techniques, which analyzes the likely current state of the atmosphere and determines the error of this analysis on the basis of incomplete and potentially faulty observations and an approximate description of the atmosphere given by the forecast model equations. The observed data used by DWD are provided by a world-wide Meteorological Observation Network (see Fig. 1.1). Considering the types of observations we can distinguish them in conventional observations (i.e., in situ observations) and non-conventional observations (i.e., remote sensing observations). The most important conventional observations are

- SYNOP data, i.e., temperature, pressure, humidity and 10-m wind at surface levels, measured by synoptic stations, at uniform times (i.e., at least at 00, 06, 12 and 18 UTC);

- SHIP data, i.e., weather reports from ships;
- DRIBU data, i.e., pressure and wind at or near sea surface level, measured by drifting buoys;
- TEMP data, i.e., highly accurate vertical profiles of temperature, humidity and wind in the upper air. The radiosondes are launched (nearly) simultaneously worldwide twice a day;
- PILOT data, i.e., wind measurements in the free atmosphere by tracking small ascending small balloons;
- PROFILERS data, i.e., measuring vertical wind profiles with remote sensing procedures, provide wind speed and wind direction observations at very high temporal and vertical resolutions;
- AIREP data (manual aircraft reports), AMDAR (Aircraft Meteorological Data Relay) and ACARS (automatic aircraft reports) supply vertical profiles of temperature and wind. AMDAR and ACARS systems usually provide more information than AIREP. During landing and take-off, ACARS deliver data in quantity, quality and location comparable to radiosondes.

Non-conventional observations are data from weather satellite and radar[1]. Weather satellite data are more and more being used because of their almost global coverage. However, accuracy and resolution of satellite data are inferior to those of radar data, so their use are of less importance for convective-scale models, especially over land, where more and more countries invest to achieve a good areal radar coverage. Modern weather radars are mostly polarimetric pulse-Doppler radars, whose typical observables are reflectivity, Doppler velocity and polarimetric parameters. Reflectivity is the amount of transmitted power returned to the radar receiver and depends among others on scatterer concentration, size, phase (liquid or ice) and orientation. Based on that, raw estimate of the precipitation rate are possible. Doppler velocity is basically a volume-averaged measure of the component of scatterer motion away from or toward a radar. It is

[1]The term RADAR was coined in 1940 by the United States Navy as an acronym for RAdio Detection And Ranging. The term radar has since entered English and other languages as the common noun radar, losing all capitalization.

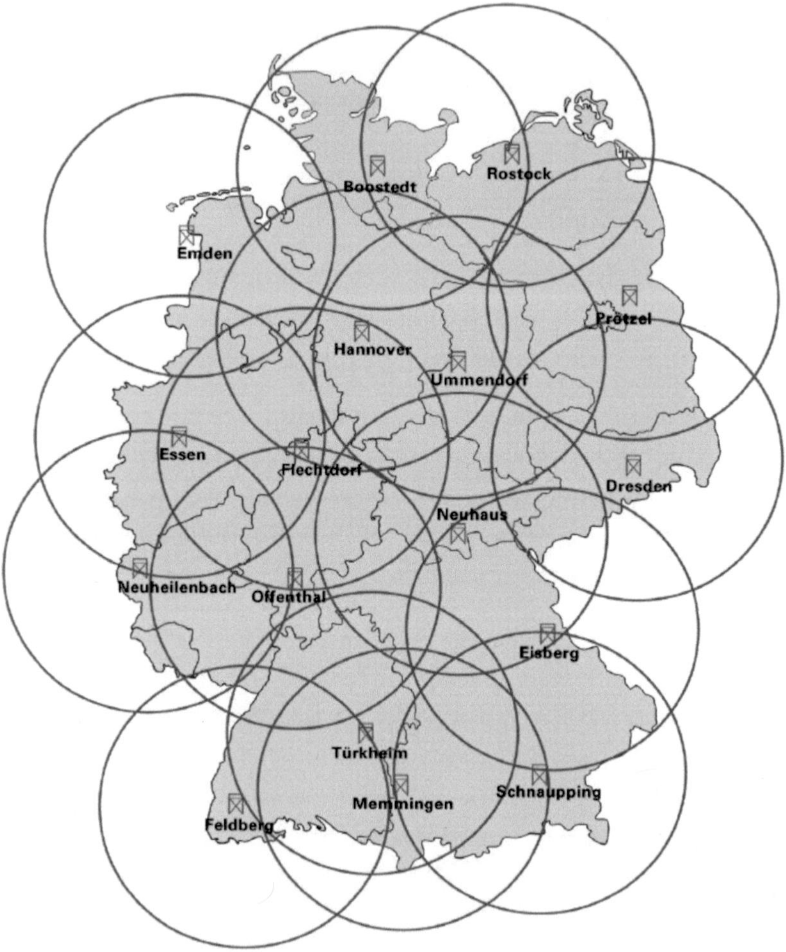

Fig. 1.2.: Map of the DWD radar network (from DWD)

deduced from Doppler phase shift from one pulse to the next and can be used to estimate the precipitation's motion. Most liquid hydrometeors have a larger horizontal axis due to the drag of air during sedimentation, which makes the water molecule dipole orient in the horizontal direction so radar pulses are generally polarized horizontally to receive the maximal power. A polarimetric radar sends out horizontally and vertically oriented pulses simultaneously or one after another and records the reflected power from both. By comparing these reflected power returns in different ways (ratios, correlations, etc.), it is possible to differentiate precipitation types (spherical drops, irregular ice particles, hail, etc.), non-meteorological targets, and to produce better rainfall estimates. However, with a single radar it is difficult to observe synoptic scale weather phenomena, such as cold fronts in the midlatitudes. This highlights an essential feature of radar observations and, in fact, that of all meteorological observations, namely the need for networking. A wide range of atmospheric phenomena can be observed with a well-distributed network of radars, which supports many applications such as operational weather monitoring and nowcasting. DWD is currently in the process of installing/upgrading a radar network of 16 dual-polarimetric C-band Doppler radar stations with a uniform scan strategy (17 after completion of network renewal). They are distributed throughout Germany for complete coverage (see Fig. 1.2 and details in Tab. A.1 in Appendix A) and delivers radar volume scans every 15 minutes for the Doppler velocity and intensity and every 5 minutes a precipitation scan, with a high spatial resolution of 1 km in range and 1° in azimuth. The Doppler volume scan (dual PRF, 800/1200 Hz, maximum range 124 km) is comprised of single sweeps with 18 elevations ranging from 0.5° to 37°. A sweep means a complete antenna revolution at a constant elevation, which geometrically corresponds to a cone. The intensity volume scan (500 Hz, max. range 256 km) has a larger areal cover and considers only the lower 5 elevations from 0.5° to 4.5°. The precipitation scan (600 Hz, max. range 150 km) is a terrain-following scan. In order to observe precipitation near the Earth's surface, the radar beam is then closely following the horizon line with an elevation offset, which should be large enough to minimize beam blockage and suppress clutter from orographic obstacles. Here, what we call "beam" is understood as a collection of neighboring rays (infinitesimally thin subparts of a beam) which individually may undergo different refraction, leading to a distortion of the beam.

However, to date only conventional data are operationally assimilated into the COSMO-model except the radar-derived precipitation rates, which are assimilated using the latent heat nudging approach (Stephan et al., 2008). Note that there are only $O(10^6)$ conventional observations available but $O(10^7)$ state variables in an NWP model. Thus, it is impossible to use those observations alone to provide the initial conditions for the model, in other words, NWP is an under-determined initial value problem. Since the weather radar is regarded as the only tool with the ability of observing the microphysical processes and dynamical movements in rapidly developing mesoscale weather phenomena on relevant temporal and spatial resolutions and COSMO-model provides the possibility to incorporate high frequency measurements into the model, the data gap left by the conventional observations could at least partially be filled by employing radar observations.

However, radar observations do not provide explicit measurements of the model variables, and it is difficult to estimate relevant model variables from radar measurements because their relations are complex and usually not unique. One possibility is to apply a so-called radar forward operator which simulates the measurement process of radar observables from the prognostic model outputs and allows for comparisons in terms of radar observables. Then, based on this operator, radar data maybe assimilated by using advanced techniques (see below). A few operators are already available in publications, which often just concentrate on specific aspects. For instance, Krajewski and Chandrasekar (1993) simulated radar reflectivity for realistic precipitation events using a stochastic space-time model and a statistically generated drop size distribution. Haase and Crewell (2000) developed a complete radar reflectivity simulator, the RADAR Simulation Model (RSM), which used the three dimensional fields of LM. A more advanced tool has been presented by Caumont (2006) at Meteo-France , who took, among others, Doppler observations and beam bending into account. SynPolRad, designed by Pfeifer et al. (2008), integrated polarimetric radar quantities using T-matrix calculations into RSM. With respect to Doppler velocity, Sun and Crook (1997) considered the vertical fall speed of hydrometeors in their operator and extended it in Sun and Crook (2001) by taking the beam broadening effect in vertical direction into account. Weighting the fall velocity by reflectivity was introduced by Wu et al. (2000) and Tong et al. (2008). A further refinement regarding weighting Doppler velocity by reflectivity can be found, for instance, in May et al. (2007), Cheong et al. (2008) as well as Caumont and Ducrocq (2008).

The goal of this thesis is to develop a comprehensive modular radar observation operator, which comprises all relevant physical aspects of radar cloud measurements in a quite accurate way, but offers also a variety of possible simplifications for each module. The choice of options has to consider a balance between physical accuracy and computational effort inasmuch as the model sophistication allows it, which may vary according to special applications, such as radar data assimilation or verification of cloud microphysical parametrizations. The latter one is done jointly with another PhD-project Jerger et al. (2012).

In terms of data assimilation, various techniques have been explored in the last few decades, which combine a prior forecast state with observations to produce an estimate of the analysis state of the atmosphere, which will be used as initial conditions for the next NWP run. The prior forecast state is also known as the background state, while the analysis state is obtained after the observational data have been assimilated into the background state. Those techniques can be divided into two main categories depending on the fact whether they are statistical methods or not.

The most important non-statistical approach is nudging (Hoke and Anthes, 1976). This method involves adding a term to the prognostic model equations that effectively nudges the solution towards observations. This nudging term is time dependent and should be large enough to be influential on the solution but small enough not to dominate the other terms (Stauffer and Seaman, 1990). Up to date, data assimilation in the COSMO-model is operationally done by nudging, but this has several disadvantages. First, it doesn't contain a mathematical formalism to determine a theoretically optimal solution to the analysis problem. Hence, there are several free parameters, whose optimal values can only be roughly estimated by means of physical reasoning and tuning experiments. Second, nudging can be only applied on the prognostic variables of models and thus its application to remote sensing data from satellites and weather radars is limited.

With regard to statistical techniques there is a mathematical representation of unknown uncertainties involved in observations and model states, so the means about which we try to find a solution is statistical (i.e., maximum likelihood, or probability). The goal is then to find an optimal combination of the model forecast background and observations with weights determined by their error statistics. It is done by minimizing a cost function, defined as a measure of the difference between the forecast states and observations. Depending on the computational methods solving this minimization problem, there are two main types of statistical techniques in modern weather forecasting (Bouttier and Courtier, 2002): variational assimilation methods and sequential methods. The former one integrates both the nonlinear model and its adjoint model over the assimilation window to compute the gradient of the cost function, and this process is repeated until a sufficient approximation to the minimum is obtained. Many meteorological centers use three-dimensional variational data assimilation (3D-VAR) due to its ease of implementation and statistical reliability. 3D-VAR assumes that within the assimilation time window all observations are all taken at the actual analysis update time and the forecast error covariance is constant. So it does not include the dynamic model in the minimization algorithm and the effects of "errors of the day", that is, having forecast error covariances reflecting the current atmospheric state. Four-dimensional variational data assimilation (4D-VAR) is a direct generalization of 3D-Var, in which the observations now are distributed in time and compared to the forecast state at the exact time, and the forecast error statistics are implicitly flow-dependent. 4D-VAR evolves the forecast error covariance implicitly from a constant initial forecast error covariance within the assimilation window and computes the model trajectory which best fits the observations distributed within the assimilation window with the dynamics described by the model. Although 4D-VAR shows especially good performance in a longer assimilation window, high costs arise from the development and maintenance of the adjoint model, which have limited its operational applications in large and complicated NWP systems.

Sequential methods include the linear Kalman filter (KF) and its nonlinear extension, the extended Kalman filter (EKF) (Gelb, 1974), which process in two steps. In the first forecast step, the model is used to integrate an earlier state estimate over the assimilation window to provide a forecast state at the time of the latest observations. In the analysis step the observations are used to improve the prior forecast state, producing a current state estimate referred to as an analysis. This analysis is used to initialize the next forecast, which is subsequently used in the next analysis, and so forth. EKF also has to use costly adjoint model. The Ensemble Kalman Filter (EnKF) is an attempt to overcome this disadvantage, which applies estimation theory with a Monte Carlo method to the conceptual and mathematical framework of KF. The starting point is to choose a set of sample points, i.e., an ensemble of state estimates that represents the initial probability distribution of the state. These sample points are then propagated through the true nonlinear system and the probability density function of the actual state is approximated by the ensemble of the estimates. EnKF retains the flow-dependent nature of the forecast error covariance matrix of KF. In the past decade, several types of EnKF have been developed. An important development was achieved by Houtekamer and Mitchell (1998), using an ensemble of pseudo-random perturbed observations to estimate the correct statistics from the analysis ensemble. A second type of EnKF is a class of deterministic (square root) filters (Anderson, 2001), which consist of a single analysis based on the ensemble mean, and the analysis perturbations are obtained from the square root of the KF analysis error covariance. Given the same size of ensemble, the square root filters are more accurate than perturbed observation filters because random errors are introduced through the perturbed observations (Whitaker and Hamill, 2002). So far different square root filters have been explored, such as Ensemble Adjustment Kalman filter (EAKF) (Anderson, 2001) and Ensemble Transform Kalman Filter (ETKF) (Bishop et al., 2001). The latter approach improves the computational efficiency of the algorithm by transforming the matrices into the subspace spanned by the forecast ensemble perturbations and doing all the matrices computation in this subspace.

It should be emphasized that a low dimensionality of the ensemble usually introduces sampling errors in the forecast error covariance, especially at long distances. Simply enhancing the size of ensemble reduce the efficiency of the method and can not be the ideal solution. Fortunately, adopting a covariance localization is able to mitigate this problem by greatly increasing the number of degrees of freedom available to fit the data. For instance, Ott et al. (2004) developed the Local Ensemble Kalman Filter (LEKF), which carries out the analyses locally in space. This reduces the computational efforts because the analyses at each grid point are independent and thus can be performed in parallel. Hunt et al. (2004) proposed an alternative type of LEKF using the ensemble transform approach introduced by Bishop et al. (2001), which is called Local Ensemble Kalman Filter (LETKF). Adding the dimension of time, Hunt et al. (2007) extended the LETKF to 4D-LETKF, which shares the main advantage of 4D-VAR to assimilate asynchronous observations at the right time.

In the field of radar data assimilation, research efforts have been achieved in the last decade, most of which have been carried out in research mode regarding specific convective case studies. Sun and Crook (1997) demonstrated a 4D-VAR scheme to assimilate radar reflectivity and Doppler velocity into a simulated moist convection case. In spite of the encouraging results, several difficulties such as the construction of the moist adjoint retrieval model and specification of background error matrix have been encountered. Lindskog et al. (2004) investigated the impact of the assimilation of Doppler velocity and velocity-azimuth display (VAD) profiles, deduced from the Swedish radar network, using the 3D-VAR of hydrostatic High-Resolution Limited Area Model (HIRLAM). Xiao and Sun (2007) introduced a radar reflectivity 3D-VAR data assimilation scheme within the fifth-generation Pennsylvania State University-National Center for Atmosphere Research (NCAR) Mesoscale Model (MM5), where some positive impacts have been found on rainfall forecasts for two particular convective cases in Korea. This approach was then operationally implemented in the Korean Meteorological Administration Doppler Radar Network in another study (Xiao et al., 2008). In recent years, EnKF has been becoming more and more appealing in radar data assimilation. The first successful study of EnKF with radar observations appeared in Snyder and Zhang (2003), where simulated radar data of single convective cells were assimilated into a cloud model. The same EnKF was tested with real radar data of a tornadic supercell in Dowell et al. (2003). The first pseudo-operationally regional-scale EnKF system was established at the University of Washington in January, using the WRF model (Torn and Hakim, 2008). Zhang et al. (2009) demonstrated that the radar data assimilation could improve the initialization and forecast of Hurricane Humberto. Overall, all of these studies show a positive impact of the assimilation of radar data on the short-range QPF. These promising results and the fact that DWD runs a radar network with a complete coverage over Germany have motivated us to take advantage of this significant amount of information to acquire more precise initial conditions of NWP, so as to improve the quality of QPF, especially at the convective scale. Since 2010, COSMO started with the priority project "KENDA" (Km-Scale Ensemble-Based Data Assimilation) under the lead of DWD, with the goal to provide suitable perturbed initial conditions for an ensemble prediction system. One of its main tasks is to develop a general 4D-LETKF data assimilation system for the COSMO-model. As part of this task, an efficient radar forward operator is desired for radar data assimilation in the new LETKF-system.

1.2. Outline

The present thesis is organized as follows. Chapter 2 briefly describes the COSMO-model. In Chapter 3, fundamentals of weather radars are described and a detailed description of the radar forward operator is given in a modular manner, followed by corresponding experiments. Chapter 4 is devoted to an overview of code implementation of the operator because it is one of major efforts in this thesis. The performance of the forward operator is assessed in two case studies in Chapter 5. Chapter 6 introduces the theory of the traditional Kalman Filter and its variants in brief. The following Chapter 7 addresses the results of our first data assimilation experiments. Finally, the last chapter gives a summary of this thesis, draws some conclusions and hints to some perspectives.

2. Description of the COSMO-model

The COSMO-model is based on the former NWP model LM (Lokal Modell). LM is a nonhydrostatic fully compressible regional atmospheric forecast model that has been developed and used at DWD and used operationally since December 1999 for both operational NWP and scientific applications on the meso-β (20-200 km, dealing with phenomena like sea breezes) and meso-γ scales (2-20 km, dealing with phenomena like thunderstorms and complex terrain flows). The basic equations of LM describe compressible flow in a moist atmosphere, which are formulated in rotated geographical coordinates with terrain following heights and consider various parametrization schemes for characterization of physical processes (see Section 2.4).

In addition to the physical basis of the forecast model itself, LM requires other components, e.g., data assimilation, interpolation of boundary conditions from a driving model, in order to run the model in NWP-mode or for case studies.

The further development in the field of high-resolution modeling has been done in close cooperation with other European weather services including those from Greece, Poland, Romania, Russia and Switzerland. In 2007, LM was renamed as COSMO to show this joint effort. Actually, DWD operationally runs two configurations. According to the configuration in which the model is run, the model name is specified by the appendix. For instance, COSMO-EU (COSMO Europe, Fig. 2.1a) covers the Eastern Atlantic and Europe with $665 \times 657 = 436905$ grid points at a horizontal resolution of 7 km and 40 vertical levels from the surface up to approximately 24 km, that is $436905 \times 40 \sim 17.5$ million grid points in total. Since 2003, DWD has been developing a new version of the COSMO-model called COSMO-DE with a horizontal resolution of 2.8 km ($\sim 0.025°$). The COSMO-DE model has been operationally run since April 2007 and provides forecasts every three hours, each forecast has a run time of 21 hours. The domain of COSMO-DE (Fig. 2.1b) covers a field of about $1300 \times 1200\,\mathrm{km}^2$ including Germany, Switzerland, Austria and some small parts of the neighbouring countries, with horizontally 421×461 grid points.

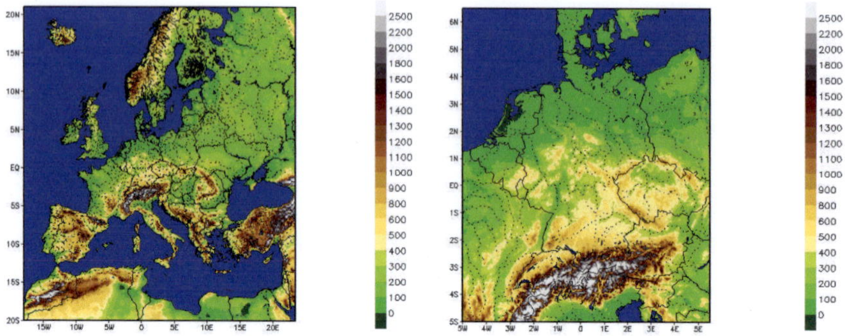

(a) The domain of COSMO-EU

(b) The domain of COSMO-DE

Fig. 2.1.: Orography (height [m] in color bar) of the operational domains of the COSMO-EU (left) and COSMO-DE (right) at DWD

The development of the COSMO-model is an ongoing task. The following sections introduce briefly the main features and characteristics. For the present thesis, the version COSMO-4.21 has been used. For a more detailed description, we refer to Doms and Schättler (2002).

2.1. Model equations

To describe the atmospheric state and its spatio-temporal development in an appropriate manner, the atmosphere is treated as a multicomponent continuum which is composed of dry air, water vapor, liquid water and water in solid state forming an ideal mixture. Water in liquid and solid forms may be further divided into various categories as cloud droplets, raindrops, pristine ice crystals, rimed aggregates of crystals, graupel and hail, etc.. Considering forces from gravity and earth rotation (by the Coriolis force) as well as internal processes due to heat, mass and momentum balances and phase changes of water, the general hydrothermodynamic equations describing compressible nonhydrostatic flow in a moist atmosphere without any scale approximations are given by:

$$\rho\frac{d\vec{v}}{dt} = -\nabla \mathsf{p} + \rho\vec{g} - 2\vec{\Omega}\times(\rho\vec{v}) - \nabla\cdot\vec{\vec{\tau}} \quad , \tag{2.1}$$

$$\frac{d\mathsf{p}}{dt} = \frac{c_\mathsf{p}}{c_\mathsf{v}}\mathsf{p}\nabla\cdot\vec{v} + \left(\frac{c_\mathsf{p}}{c_\mathsf{v}} - 1\right)Q_h + \left(\frac{c_\mathsf{p}}{c_\mathsf{v}}\right)Q_m \quad , \tag{2.2}$$

$$\rho c_\mathsf{p}\frac{d\mathsf{T}}{dt} = \frac{d\mathsf{p}}{dt} + Q_h \quad , \tag{2.3}$$

$$\rho\frac{dq^x}{dt} = -\nabla\cdot\vec{J}^x + I^x \quad , \tag{2.4}$$

$$\rho = \mathsf{p}/\left[R_d(1+\alpha^m)\mathsf{T}\right] \quad . \tag{2.5}$$

The index x represents a specific constituent of the mixture with

$x = d$	for dry air,
$x = v$	for water vapor,
$x = l$	for liquid water, and
$x = f$	for water in the solid state, i.e., ice.

The total derivative of a field ψ is related to partial time and space variations by the Euler decomposition:

$$\frac{d\psi}{dt} = \frac{\partial\psi}{\partial t} + \vec{v}\cdot\nabla\psi \quad . \tag{2.6}$$

The list of symbols in Eqs. (2.1-2.5) are given in Tab. 2.1.

Several modifications should be made in Eqs. (2.1-2.5) since they are numerically solved on a structured grid. First of all, differential operators appearing in the equations are approximated by difference operators and thus only valid in the limit when the time interval Δt and the spatial increment $\Delta\mathsf{V}$ ($\Delta\mathsf{V} := \Delta x\Delta y\Delta z$) approaches zero. For a physically meaningful interpretation, on one hand, spatial increment $\Delta\mathsf{V}$ muss be much larger than the spacing between molecules to contain a sufficient number of molecules to apply statistical thermodynamics, but on the other hand, it must be much smaller than macroscopic dimensions so that the values of variables do not strongly change within Δt and $\Delta\mathsf{V}$. These conditions restrict the direct application of Eqs. (2.1-2.5) to space scales on the order of about 1 cm and to time scales of about 1 s.

t	Time
p	Pressure
T	Temperature
ρ_x	Partial density of mixture constituent x
$\rho = \sum_x \rho_x$	Total density of the air mixture
$q^x = \rho^x/\rho$	Mass fraction (specific content) of constituent x
$\vec{v} = (u, v, w)$	Wind vector
c_p, c_v	Heat capacities for constant pressure and volume
$\alpha^m = \left(\frac{R_v}{R_d} - 1\right) q^v - q^l - q^f$	Moisture term
R_v, R_d	Gas constant for water vapor and dry air
I^x	Source/sinks of constituent x
$\vec{J}^x$	Diffusion flux of constituent x
Q_h	Diabatic heating
$Q_m = \rho R_d \mathrm{T} \frac{d\alpha^m}{dt}$	Impact of changes of humidity
$\vec{\vec{\tau}}$	Stress tensor due to friction
$\vec{\Omega}$	Constant angular velocity vector of earth rotation
$\vec{g}$	Apparent acceleration of gravity
∇	Gradient Nabla operator

Tab. 2.1.: The list of symbols in Eqs. (2.1-2.5)

However, as mesoscale circulations have horizontal scales up to 100 km and vertical scales up to 10 km, using a numerical model of grid spacing on the order of 1 cm to simulate such flows is not feasible. To circumvent this problem, the equations are averaged over specified space and time step, and then the meteorological variables are split up into a mean value (the grid scale value) and its deviation (the subgrid scale value). Furthermore, some assumptions are made to simplify the equations (Doms and Schättler, 2002):

1. all molecular fluxes are replaced by turbulent fluxes formally written in the same manner. Components of the turbulent stress tensor follow from a parametrization of turbulence.

2. the specific heat of moist air is replaced by that of dry air,

3. the diabatic terms Q_h and Q_m are neglected,

4. temperature changes due to buoyant heat and moisture fluxes are neglected.

These assumptions are typically justified with the fact that the atmosphere air can be described as a very diluted mixture with respect to the water constituents. Moreover, in order to enhance the numerical accuracy, the thermodynamic variables temperature T, pressure p and density ρ can be formally expressed as the sum of a height dependent base state as reference value (indicated with subscript 0) and a space and time dependent deviation (indicated by a prime):

$$\mathsf{T} = \mathsf{T}_0(z) + \mathsf{T}' \quad ,$$
$$\mathsf{p} = \mathsf{p}_0(z) + \mathsf{p}' \quad ,$$
$$\rho = \rho_0(z) + \rho' \quad .$$

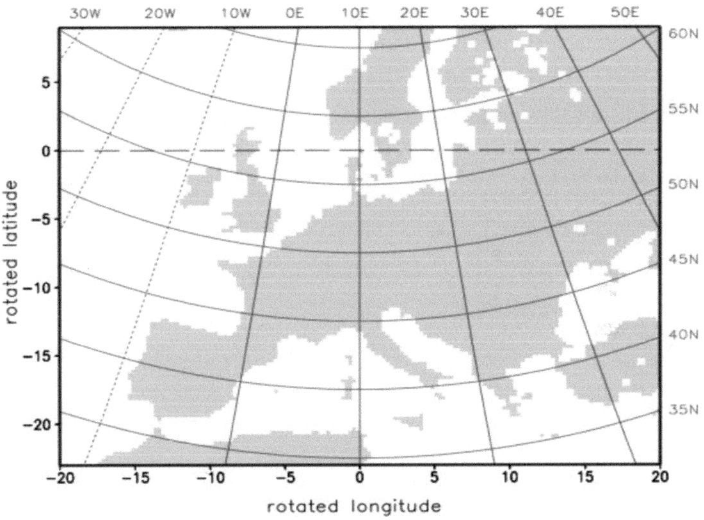

Fig. 2.2.: Geographical longitude and latitude in the unrotated grid. The dashed line indicates the equator in the rotated grid with pole coordinates 32.5°S and 10.0°E in the unrotated system. The rotated 0° meridian corresponds to the 10°E geographical meridian (Doms and Schättler, 2002).

To take the (nearly) spherical nature of the earth into account, the usual way would be to transform the model equations into geographical coordinates. However, regarding practical applications on a large model domain, such spherical coordinates cause numerical problems arising from the convergence of the meridians and the resulting pole singularities. A suitable way to minimize these problems is to introduce rotated spherical coordinates (λ, φ, z), where λ is geographical rotated longitude, φ is geographical rotated latitude, z is geographical height above mean sea level. It is done by shifting the North Pole of the new system in such a way that the intersection of the equator and the prime meridian of the new system passes through the centre of the model domain and thus the convergence of the meridians can be minimized (Fig. 2.2). Transformation equations can be found in Doms and Schättler (2002).

In spherical coordinates, the vertical coordinate z is curvilinear but orthogonal. When surface terrain is considered, it becomes very complicated to formulate the lower boundary conditions and quite expensive to find the numerical solution of the basic equations. An elegant way to alleviate this problem is the transformation of z to a terrain-following coordinate system, where the lowest surface of constant vertical coordinate becomes conformal to the terrain height. The new vertical coordinate ζ is a time-independent function of λ, φ and z. This is different from the pressure based coordinate system of most hydrostatic models, where the surfaces of constant vertical coordinate move in space with changing surface pressure. Fig. 2.3 views a sketch of terrain-following coordinate system over orography.

In order to keep the numerical formulation of the model equations independent from the choice on ζ, the coordinate transformation will be done in two steps. The first step involves a terrain-following transformation using a user-specified coordinate $\tilde{\zeta}$. In the second step, $\tilde{\zeta}$ is mapped to the computational coordinate ζ by a monotonic function m in the form $\tilde{\zeta} = m(\zeta)$. Since m can be any monotonic function, we define this function to map (by its inverse) the coordinate $\tilde{\zeta}$ to the index space with top-down increasing indices and an equidistant grid spacing of $\Delta\zeta = 1$. Fig. 2.4 illustrates this two-step transformation. In fact, there are three options for the terrain-following $\tilde{\zeta}$ in the COSMO-model. The first one is a reference-pressure based coordinate, the second one is a Gal-Chen height-based coordinate (Gal-chen and Somerville, 1975) and the third one is the height-based SLEVE (Smooth Level VErtical) coordinate according to Schär et al. (2002).

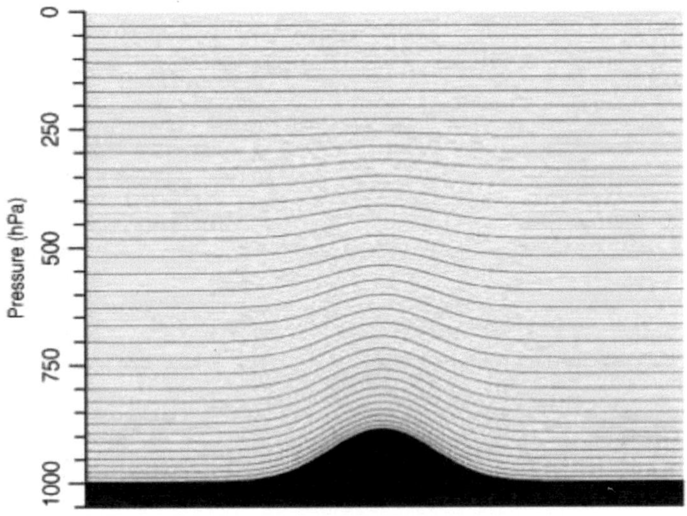

Fig. 2.3.: Sketch of a terrain following coordinate system (Doms and Schättler, 2002)

Introducing these modifications into Eqs. (2.1-2.5), we obtain the final version of model equations including seven prognostic equations for horizontal wind velocity, vertical wind velocity, perturbation pressure, temperature, water vapor, liquid and solid forms of water and one diagnostic equation for total density of air (Doms and Schättler, 2002):

- Horizontal wind velocity components

$$\frac{\partial u}{\partial t} = -\left\{\frac{1}{R_E \cos\varphi}\frac{\partial E_h}{\partial \lambda} - vV_a\right\} - \dot{\zeta}\frac{\partial u}{\partial \zeta} - \frac{1}{\rho R_E \cos\varphi}\left(\frac{\partial p'}{\partial \lambda} - \frac{1}{\sqrt{\gamma}} - \frac{1}{\sqrt{\gamma}}\frac{\partial p_0}{\partial \lambda}\frac{\partial p'}{\partial \zeta}\right) + M_u \quad , \tag{2.7}$$

$$\frac{\partial v}{\partial t} = -\left\{\frac{1}{R_E}\frac{\partial E_h}{\partial \varphi} - uV_a\right\} - \dot{\zeta}\frac{\partial v}{\partial \zeta} - \frac{1}{\rho R_E}\left(\frac{\partial p'}{\partial \varphi} - \frac{1}{\sqrt{\gamma}} - \frac{1}{\sqrt{\gamma}}\frac{\partial p_0}{\partial \varphi}\frac{\partial p'}{\partial \zeta}\right) + M_v \quad , \tag{2.8}$$

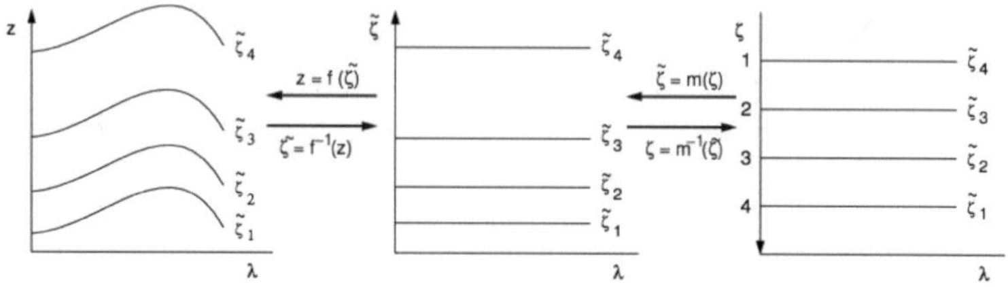

Fig. 2.4.: Mapping of an irregular curvilinear grid with a terrain following coordinate $\tilde{\zeta}$ onto a rectangular equidistant grid ζ labeled by integers (Doms and Schättler, 2002)

- Vertical wind velocity

$$\frac{\partial w}{\partial t} = -\left\{\frac{1}{R_E \cos\varphi}\left(u\frac{\partial w}{\partial \lambda} + v\cos\varphi\frac{\partial w}{\partial \varphi}\right)\right\} - \dot{\zeta}\frac{\partial w}{\partial \zeta} - \frac{g}{\sqrt{\gamma}}\frac{\mathrm{p}_0}{\rho}\frac{\partial \mathrm{p}'}{\partial \zeta} + M_w$$
$$+ g\frac{\rho_0}{\rho}\left\{\frac{\mathrm{T}-\mathrm{T}_0}{\mathrm{T}} - \frac{\mathrm{T}_0\mathrm{p}'}{\mathrm{T}\mathrm{p}_0} + \left(\frac{R_v}{R_d} - 1\right)q^v - q^l - q^f\right\} \quad , \qquad (2.9)$$

- Perturbation pressure

$$\frac{\partial \mathrm{p}'}{\partial t} = -\left\{\frac{1}{R_E \cos\varphi}\left(u\frac{\partial \mathrm{p}'}{\partial \lambda} + v\cos\varphi\frac{\partial \mathrm{p}'}{\partial \varphi}\right)\right\} - \dot{\zeta}\frac{\partial \mathrm{p}'}{\partial \zeta} + g\rho_0 w - \frac{c_\mathrm{p}}{c_\mathrm{v}}\mathrm{pD} \quad , \qquad (2.10)$$

- Temperature

$$\frac{\partial \mathrm{T}}{\partial t} = -\left\{\frac{1}{R_E \cos\varphi}\left(u\frac{\partial \mathrm{T}}{\partial \lambda} + v\cos\varphi\frac{\partial \mathrm{T}}{\partial \varphi}\right)\right\} - \dot{\zeta}\frac{\partial \mathrm{T}}{\partial \zeta} - \frac{1}{\rho c_\mathrm{d}}\mathrm{pD} + Q_\mathrm{T} \quad , \qquad (2.11)$$

- Water vapor

$$\frac{\partial q^v}{\partial t} = -\left\{\frac{1}{R_E \cos\varphi}\left(u\frac{\partial q^v}{\partial \lambda} + v\cos\varphi\frac{\partial q^v}{\partial \varphi}\right)\right\} - \dot{\zeta}\frac{\partial q^v}{\partial \zeta} - \left(S^l + S^f\right) + M_{q^v} \quad , \qquad (2.12)$$

- Liquid and solid forms of water

$$\frac{\partial q^{l,f}}{\partial t} = -\left\{\frac{1}{R_E \cos\varphi}\left(u\frac{\partial q^{l,f}}{\partial \lambda} + v\cos\varphi\frac{\partial q^{l,f}}{\partial \varphi}\right)\right\}$$
$$-\dot{\zeta}\frac{\partial q^{l,f}}{\partial \zeta} - \frac{g}{\sqrt{\gamma}}\frac{\rho_0}{\rho}\frac{\partial P_{l,f}}{\partial \zeta} + S^{l,f} + M_{q^{l,f}} \quad , \qquad (2.13)$$

- Total density of air (cf. Eq. (2.5)

$$\rho = \mathrm{p}\left\{R_d\left(1 + \left(\frac{R_v}{R_d} - 1\right)q^v - q^l - q^f\right)\mathrm{T}\right\}^{-1} \quad , \qquad (2.14)$$

where g is the gravity acceleration, D is divergence of the wind field, $E_h = \sqrt{u^2 + v^2}$ is kinetic energy of horizontal motion, $\sqrt{\gamma}$ is variation of reference pressure with ζ, V_a is the vertical component of the absolute vorticity, $P_{l,f}$ are precipitation fluxes, the terms M_ψ denote contribution from subgrid scale processes as, e.g., turbulence and convection. Q_T summarizes the diabatic heating rate due to this processes. The various cloud microphysical sources and sinks due to phase changes are denoted by S^l and S^f. The calculation of all these terms related to subgrid-scale processes is done by physical parametrization schemes. Notice that the pressure equation (2.2) has been replaced by an equation for pressure deviation p'. Also note that E_h and V_a enable an elegant formulation of horizontal advection, Coriolis force and an Earth's curvature term together. However, the numerical discretization is only done using the original u and v.

2.2. Discretized form of the model equations

For the numerical solution of the continuous model equations listed in the previous section, spatial and temporal discretization of the equations must be done. The spatial discretization is realized by model grid structure and for the temporal discretization a Runge-Kutta scheme is currently used.

2.2.1. Model grid structure

Eqs. (2.8-2.13) have been written in a terrain-following coordinate system using a generalized vertical coordinate ζ. This general form of the transformation is employed to map the irregular curvilinear grid associated with the terrain-following system $\tilde{\zeta}$ in physical space onto a regular computational grid, which is set up by constant increments:

$$
\begin{aligned}
\Delta\lambda &: \text{grid-spacing in } \lambda\text{-direction,} \\
\Delta\varphi &: \text{grid-spacing in } \varphi\text{-direction,} \\
\Delta\zeta &: \text{grid-spacing in } \zeta\text{-direction.}
\end{aligned}
$$

The computational $(\lambda, \varphi, \zeta)$-space is then represented by a finite number of grid points with integer values (i, j, k), where i corresponds to the λ-direction, j to the φ-direction and k to the ζ-direction. The position of the grid points in the computational space is defined by

$$\lambda_i = \lambda_0 + (i-1)\Delta\lambda \quad , \qquad i = 1, \ldots, N_\lambda, \tag{2.15}$$

$$\varphi_j = \varphi_0 + (j-1)\Delta\varphi \quad , \qquad j = 1, \ldots, N_\varphi, \tag{2.16}$$

$$\zeta_k = k \quad , \qquad k = 1, \ldots, N_\zeta, \tag{2.17}$$

where N_λ, N_φ and N_ζ denote the number of grid points in λ, φ and ζ-directions, respectively. Thus, λ_0 and φ_0 are the southwestern corner of the model domain in the rotated spherical coordinates (λ, φ). Every grid point (i, j, k) represents the centre of an elementary rectangular grid volume with side lengths $\Delta\lambda$, $\Delta\varphi$ and $\Delta\zeta$. Horizontally, the grid-box faces are located halfway between the grid points in the corresponding directions, i.e., at $\lambda_i \pm 1/2$, $\varphi_j \pm 1/2$ and $\zeta_k \pm 1/2$. Vertically, the grid-box faces are usually referred to as the half levels. These interfacial levels separate the model layers from each other. The model layers labeled by integers k are also referred to as main levels. The top boundary of the model domain is defined to be the half level $(\zeta = 1/2)$ above the uppermost model layer $(\zeta = 1)$ and the ζ-coordinate surface becomes conformal to the orography at the lower boundary. The half level $(\zeta = N_\zeta + 1/2)$ below the first model layer above the ground $(\zeta = N_\zeta)$ defines the lower boundary of the model.

The scalar model variables (temperature, pressure and humidity variables) are defined at the centre of a grid box (main level) while the wind components *u*, *v* and *w* are defined on the corresponding box faces (half level in the vertical) (Fig. 2.5). This grid structure is called staggered Arakawa-C/Lorenz grid and is advantageous for the discretization of the flow equations.

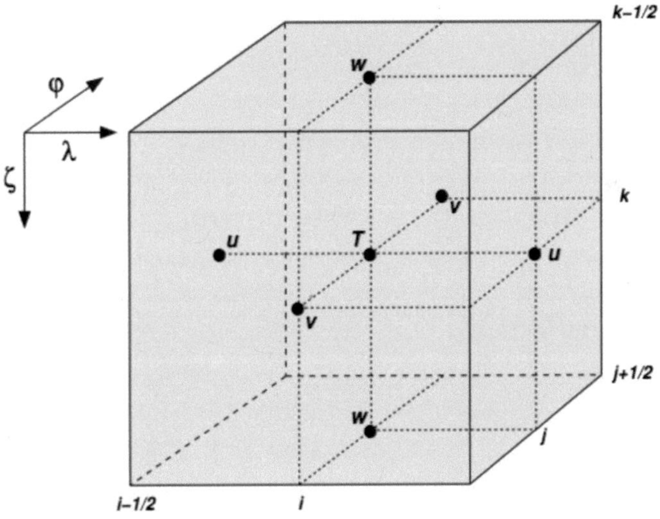

Fig. 2.5.: A grid box volume showing the Arakawa-C/Lorenz staggering of the dependent model variables (Doms and Schättler, 2002)

2.2.2. Time integration scheme

Three different time integration schemes have been implemented within the COSMO-model: the leapfrog-scheme (Klemp and Wilhelmson, 1978) and the semi-implicit solver (Thomas et al., 2000) and the third order Runge-Kutta method (RK3) of Wicker and Skamarock (2002). All these schemes use time splitting techniques, which separate the prognostic equations in terms of fast processes related to acoustic wave modes and terms in conjunction with comparatively slowly varying modes of motion (e.g., advection), but in different manners (Doms and Schättler, 2002). As RK3 is numerically quite stable and efficient, it has been operationally run at DWD since April 2007.

2.3. Initial and boundary conditions

In a limited area model as COSMO, only the lower boundary is physical due to the Earth's surface. The top and lateral boundaries are usually artificial, and have to be specified. It is important to use open or inflow-outflow lateral boundary conditions to allow the atmosphere in the model interior domain to interact with the external environment, when it comes to the simulation of real data cases or NWP purposes (Davies, 1976). Alternatively, we can use open and periodic boundary conditions for specific scientific applications (i.e., idealized simulations).

For operational applications and real data simulations, the initial conditions for the COSMO-model can be specified by forecasts interpolated from various global models, e.g., the GME model of DWD or the global IFS model from ECMWF. Alternatively, the analysis results of data assimilation in the COSMO-model can be used for initial conditions. For the purpose of idealized case studies, user-specified artificial initial data can also be chosen.

The approaches to establish the lateral boundary conditions are almost the same as for the initial conditions.

For NWP purposes, information on the variables at the lateral boundaries and their time evolution is obtained by interpolation from larger models. The use of a model at a coarser resolution for driving a high-resolution limited area model causes numerical problems, since the time evolution of the model variables is based on a set of equations differing from those of the driving model. The problems are related to a non-unique information transfer between the models at the boundaries, due to differences in the spatial resolution and model equations. These numerical noises can propagate from the lateral boundaries to the interior of the model domain. To fix this problem, a relaxation zone close to the boundaries is used, in which the variables of the high- resolution model are gradually modified to blend them with the driving model variables, and the influence of the driving model decreases exponentially with increasing distance to the domain boundary (Davies, 1976).

For idealized runs, the periodic boundary condition assumes indefinite repetition of the solution of the model equations outside the computational domain, so that the solution at a certain distance to the west (north) of the computational domain western (northern) boundary is equal to that at the same distance to the west (north) of the eastern (southern) boundary.

The top boundary of the model domain is defined as the top half level with constant computational coordinate $\zeta = 1/2$ with fixed height above mean sea level and a rigid lid, i.e., $\dot{\zeta} = 0$. For the horizontal wind velocity, the temperature and the water substances, the free-slip condition is assumed, i.e., no mass transfer across the upper boundary.

Additionally, a so-called Rayleigh damping scheme may be applied to a number of model layers just below the upper boundary to absorb upward propagating wave disturbances and to suppress gravity wave reflection at the top boundary resulting from the rigid lid upper boundary condition. The prevention of wave energy reflection at the upper boundary is vital for a correct simulation of orographically induced flows.

2.4. Physical parametrization

Atmospheric processes span horizontal scales from molecular to planetary, and they vary in time scales from less than seconds to longer than annual scales. Because of the limited spatial and temporal resolutions of atmospherical models, an important part of these physical processes is not accounted for by the explicit solution in the model grid of the basic equations. On one hand, this concerns all molecular processes as radiation, cloud microphysics and laminar transport in the immediate vicinity of solid boundaries. On the other hand, there are processes as turbulence and convection. All processes that are not explicitly simulated by the model bu considered to be important for the model results have to be treated in a special manner called parametrization. This section summarizes shortly the parametrization schemes used in the COSMO-model more details can be found in Doms and Schättler (2002).

Subgrid-scale turbulence: a prognostic equation for TKE (turbulent kinetic energy), which is a level 2.5 closure scheme (Raschendorfer, 2001), is used. A parametrization of the pressure transport term is considered in the TKE-equation, which accounts for TKE-production by subgrid thermal circulations. The whole scheme is formulated with conservative thermodynamic variables together with a statistical cloud scheme in order to consider effects from subgrid-scale condensation along the lines of Sommeria and Deardorff (1977).

Surface layer parametrization: the surface layer scheme extends the TKE-equation to the constant flux layer and introduces an additional laminar layer just above the surface. This makes it possible to discriminate between values of model variables at the rigid surface (e.g., radiative surface temperatures) and values at the roughness height (lower boundary of the turbulent atmosphere).

Grid-scale clouds and precipitation: the standard COSMO-model uses the so-called one-moment scheme for cloud physics, i.e., only the bulk masses of five or six different hydrometeor classes (water vapor, cloud water, cloud ice, snow and optionally graupel) are predicted at each grid point, by assuming a prescribed size distribution of the particles (cloud water: no assumption necessary in the simple parametrization framework; cloud ice: monodispers; rain: Gamma-distribution; snow: implicit shape based on Field et al. (2007); graupel: exponential distribution). It considers explicitly processes of cloud and ice nucleation, diffusional growth of water and ice phase, drop to drop, drop to ice, and ice to ice collision, ice multiplication, break up of raindrops, freezing and melting. A more advanced approach, a two-moment scheme, extends the description with hail and additional prognostic variables, the hydrometeor number densities (Seifert and Beheng, 2006; Seifert et al., 2006; Blahak, 2008b; Noppel et al., 2010). This allows a better parametrization of the size distribution function. But owing to their enormous computational complexity, it is currently not used for operational applications but for research. Note that without explicit statement all model runs are performed with the so-called one-moment graupel scheme (Reinhardt and Seifert, 2006) in this thesis.

Fractional cloud cover: in the parametrization schemes for grid-scale clouds and precipitation, the condensation rate for cloud water is based on saturation equilibrium with respect to water. Therefore, a grid element is either fully filled with clouds at water saturation where $q^c > 0$ (relative humidity $= 100\%$) or it is cloud free at water subsaturation where $q^c = 0$ (relative humidity $< 100\%$), so the area fraction of a grid element covered with grid-scale clouds is either 1 or 0. However, with respect to the calculation of radiative transfer and weather interpretation in postprocessing routines, it is meaningful to define a fractional cloud cover for those grid boxes where the relative humidity is less than 100% and no grid-scale cloud water exists. The calculation of the fractional cloud cover in each model layer is determined by an empirical function of relative humidity, height of the model layer, convective activity and stability (Sommeria and Deardorff, 1977).

Moist convection: either the mass-flux convection scheme with equilibrium closure based on moisture convergence (Tiedtke, 1989) or the mass-flux convection scheme with non-equilibrium CAPE-type closure (Kain and Fritsch, 1993) can be used. In addition to the closure, they differ from each other mainly in the triggering criteria for convection and the processes influencing detrainment and entrainment (Smoydzin, 2004).

Radiation: it is parameterized according to a so-called two-stream scheme of Ritter and Geleyn (1992) which considers three short wave (solar) and five long wave (thermal) spectral intervals. Clouds, aerosol, water vapor and other gaseous tracers are treated as optically active constituents of the atmosphere, which modify the radiative fluxes by absorption, emission and scattering. As an extension to the original scheme, a new treatment of the optical properties of ice particles has been introduced which allows a direct cloud-radiative feedback with the predicted ice and water content.

Soil model: A simple two-layer soil model (Jacobsen and Heise, 1982) employing the extended force-restore method is applied; snow and interception storage are included. Optionally, a new multilayer version of the model based on the direct numerical solution of the heat conduction equation can be used.

Terrain and surface data: for data like orography, land-sea mask, soil type and vegetation cover (the so-called "external" data), the model employs standard data sets provided by various sources (e.g., orography from the Global Land One-km Base Elevation (GLOBE), harmonized world soil database form the Food and Agriculture Organization of the United Nations (FAO), etc.). There is a software package at DWD, which is able to aggregate/interpolate/convert these data sets to the required external datasets for the COSMO-model (see webpage `http://www.cosmo-model.org/content/model/modules/externalParams/default.htm`). Data sets are available for different horizontal resolutions and pre-defined regions covering Europe. Other data sets can be created by DWD on request.

3. Radar forward operator

As mentioned in Section 2.4, moist convection has to be parameterized in NWP models including the COSMO-model. The inherent limitations of this subgrid parameterization usually lead to the low confidence of QPFs (Fritsch and Carbone, 2004). As pointed out in Ducrocq et al. (2000) and Trier and Manning (2004), improvements in QPFs can be achieved by both explicit treatment of moist convection and advanced parametrization of microphyiscal processes. These studies also expressed the improvement possibility by assimilating high-resolution observations such as weather radar data. As noticed, the DWD radar network provides a huge amount of data in high spatial and temporal resolutions, covering the entire COSMO-DE domain, which has encouraged the use of radar data to enhance the quality of operational analyses and forecasts.

However, a viable radar data assimilation scheme requires a tool that establishes a link between the model data and radar data and allow for a direct comparison between them. For this purpose, a model-to-observation method, the so-called "forward operator" has been developed which transforms the model outputs into radar observations and performs comparisons in terms of observed quantities. This transformation succeeds by simulating the main processes relevant to radar measurements. In the first step, radar observables are computed from predicted bulk water quantities. In the second step, the beam propagation is simulated under consideration of physical processes influencing radar measurements like attenuation, beam bending and broadening. Note that simulations have to follow the model assumptions as closely as possible for the sake of consistency. One important criterion of the operator design is modularity, which means that the operator should be comprised of several building modules and each module describes a particular physical process in radar measurements and offers several options associated with different accuracy and complexity in simulations. These options can be flexibly switched on/off in accordance with the user's needs: the radar forward operator can be used as a model evaluation tool and also as an observation operator for (operational) assimilation systems. The former one demands relative high accuracy rather than efficiency of simulations although for the latter one special emphasis must be given to simulation time, especially for operational applications. Since the development of the radar forward operator for operational data assimilation is the main objective of this thesis, it is proposed here that sensitivity experiments be performed systematically for each module of the operator (see this chapter and more in Chapter 5) and based on the results of experiments we are able to recommend how to configure the operator for the purpose of data assimilation. Moreover, the radar forward operator should be flexible enough to be able to simulate radars that operate at different frequencies and scan strategies in either research or operational modes.

Owing to the limited time given for this thesis, the polarization parameters are not considered yet, but integration of those parameters should mostly be straight-forward and we intend to implement a code for one-layered spheroidal particles in near future according to Pfeifer et al. (2008). Indeed, this will probably drastically increase the computational expense, so it is necessary to simplify the computations, e.g., by means of lookup tables, which cover the relevant range of the basic parameters and depend on assumptions about canting angle distributions and axis ratios of the spheroids as function of size.

This chapter gives at first fundamentals of weather radars and then Sections 3.6-3.8 are devoted to a comprehensive description of the radar forward operator in a module-wise way, where sensitivity experiments for beam bending and broadening are done and their outcomes are discussed. For brevity in the remainder of the work, the term "operator" refers to the radar forward operator without explicit statement.

3.1. Basic radar terms

Most weather Doppler radars are pulsed radars, which emit microwave energy from a transmitter into the atmosphere in a rapid succession of short (i.e., from tens of nanosecs to tens of microsecs) pulses (see Fig. 3.1). During the time between two transmitted pulses, the radar switches to receive mode. When these pulses impinge on objects in the atmosphere such as raindrops, hail stones, snowflakes, cloud droplets, birds, insects, dust particles, vegetation and even the ground, part of the energy bounces back towards the radar. A receiver on the radar then collects the reflected energy and

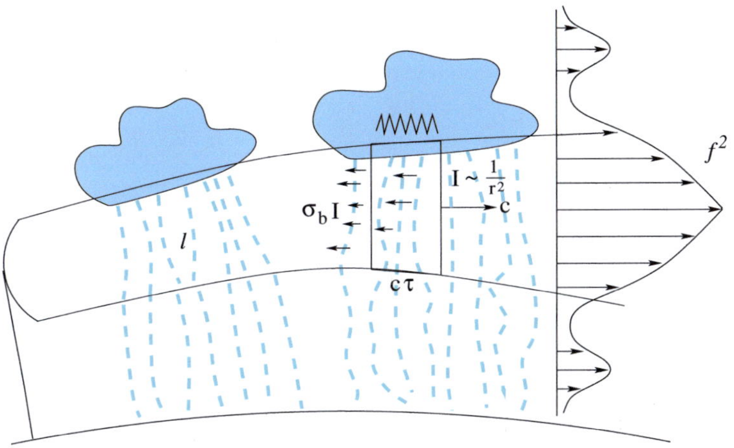

Fig. 3.1.: Schematic representation of the radar measurements: The radar antenna transmits an electromagnetic pulse that travels with light speed c through the atmosphere until it encounters scatters, in the atmosphere mostly hydrometeors. A part of energy ($\sigma_b \mathtt{I}$) will be then backscattered to the antenna (short arrows). In addition, some energy will be lost on its way due to attenuation (ℓ). τ is the pulse duration and $c\tau$ is the pulse length.

stores the data for visualization (see Fig. 3.2). The PPI (Plan Position Indicator, see Figs. 3.2a and 3.2b) is the most common type of radar display, which exhibits radar data horizontally using a map projection. In PPI mode, the radar performs a 360-degree sweep with the antenna at a specific elevation. Other radar images include MAXCAPPI (Maximum Constant Altitude Plan Position Indicator: vertical maxima projected on the horizontal plan and maxima from each level horizontally projected from south to north and from west to east, see Fig. 3.2c), SRI (Surface Rain Intensity, see Fig. 3.2d), etc..

With each pulse, a radar resolution volume, the so-called the pulse volume, positioned by its center $\vec{r_0} = (r_0, \alpha_0, \varepsilon_0)$, is illuminated with the following illumination function:

$$\mathtt{I}(\vec{r}) = C_r \frac{f^4(\phi,\theta)}{r^2} |W(r_0 - r)|^2 \quad , \tag{3.1}$$

where $\mathtt{I}$ is the emitted energy intensity, r_0 is the radial distance (also called range) between the antenna and the center of the pulse volume, r is the radial distance from antenna to an arbitrary position within the pulse volume, α_0 and ε_0 are azimuth and

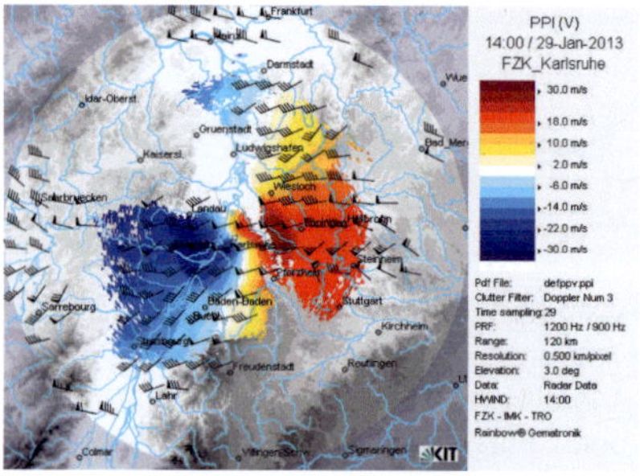

(a) PPI for Doppler velocity

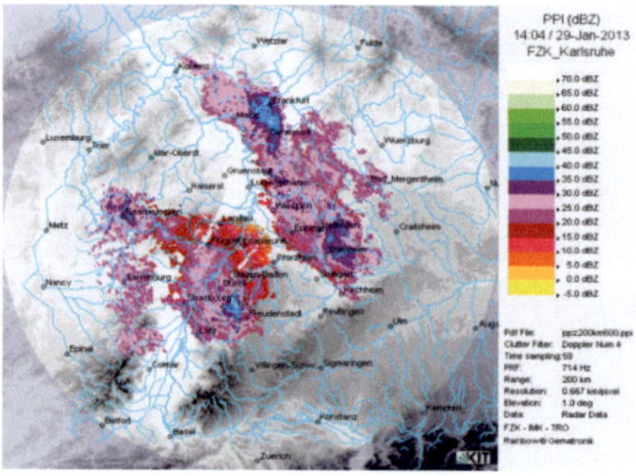

(b) PPI for reflectivity

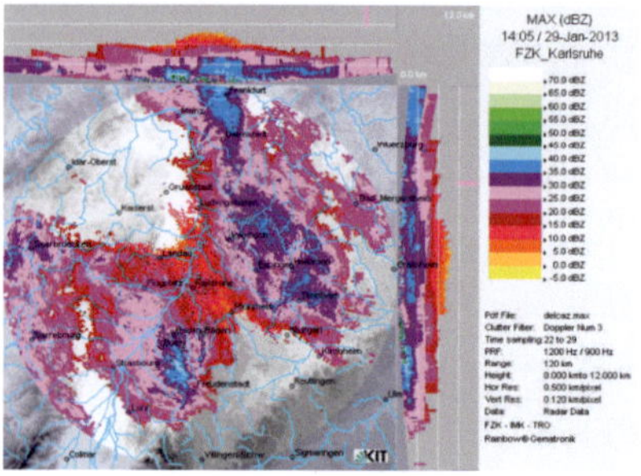

(c) MAXCAPPI

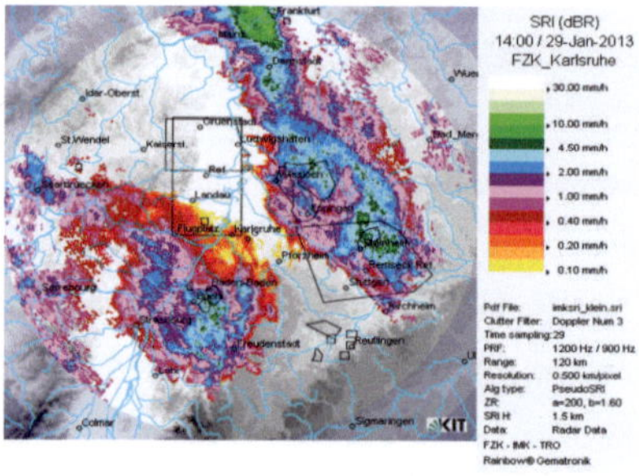

(d) SRI

Fig. 3.2.: Examples of radar displays: observations of C-Band Doppler radar at KIT Campus North on 29 January 2013

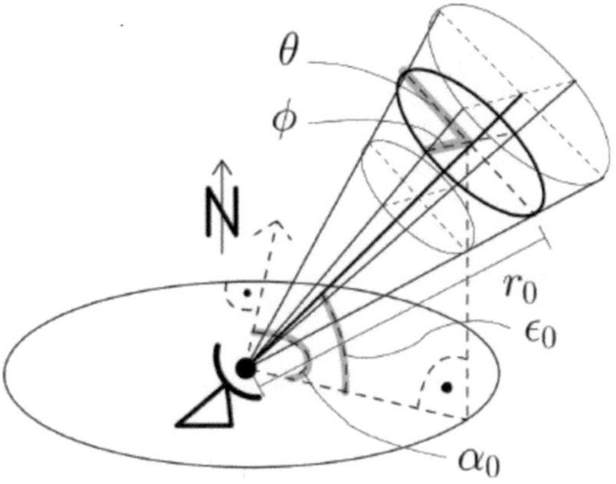

Fig. 3.3.: A single radar beam, described in the radar system $(r_0, \alpha_0, \varepsilon_0)$ and beam system (r_0, ϕ, θ). Pulse volume is represented by thin ellipses. The general radar system coordinates (r, α, ε) are determined relative to the coordinates $(r_0, \alpha_0, \varepsilon_0)$ (Blahak, 2008a).

elevation of the antenna, respectively. The coefficient C_r is the so-called radar constant and depends on radar system parameters including power transmitted P_t, antenna gain G_0 and radar wavelength λ.

The geometric dependency of quantities determining $\mathtt{I}$ is expressed in terms of the so-called "beam system" $\vec{r} = (r, \phi, \theta)$, where ϕ and θ are horizontal and vertical angles relative to the ray in the beam center, respectively. This is different from the "radar system" $\vec{r} = (r, \alpha, \varepsilon)$. The contrast of both systems is illustrated in Fig. 3.3 for a single radar beam.

The term of $1/r^2$ in Eq. (3.1) indicates that the energy intensity $\mathtt{I}$ decreases with distance by $1/r^2$ as for spherical waves. Consequently, the targets at long ranges are poorly illuminated and their echoes might be too faint to be detected, so that light precipitation becomes undetectable at long ranges. The range r is determined by measuring the delay Δt between transmission of a pulse and its echo (see Fig. 3.4), that is, $r = c\Delta t/2$, where the factor $1/2$ accounts for the two-way time delay.

The pulse duration τ is the time over which a pulse lasts (see Fig. 3.4). τ can be multiplied by the light speed to determine the pulse length ($= c\tau$). The Pulse Repetition Frequency (PRF) is the number of pulses that are transmitted per second. The reciprocal of PRF is called the Pulse Repetition Time (PRT), which is the time interval

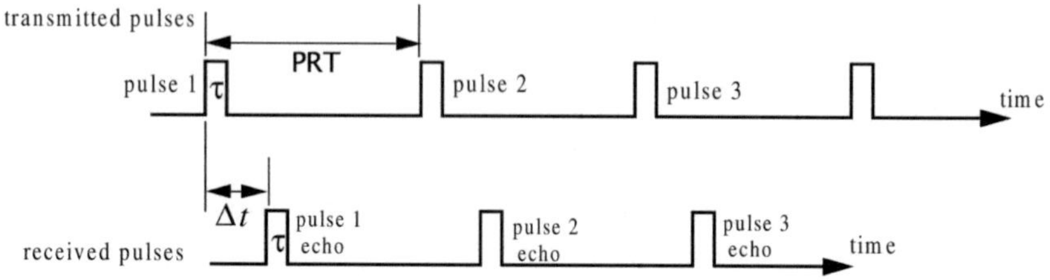

Fig. 3.4.: Train of transmitted and received pulses (Mahafza, 2000)

between the start of two consecutive pulses. When multiple pulses are transmitted, there is the possibility of a range ambiguity. To determines the range r unambiguously requires that the time interval between pulses PRT must be larger than the time for a single pulse to propagate to range r and back, that is PRT $\geq 2r/c$, and the unambiguous range is r_{max}:

$$r_{max} = \frac{c \cdot \mathrm{PRT}}{2} = \frac{c}{2\mathrm{PRF}} \quad . \tag{3.2}$$

The radar pulse volume is spacious and keeps broadening as the pulse propagates away from the radar antenna. The function $f^2(\phi,\theta)$ is the beam weighting function describing the weight at which local reflectivity and attenuation contribute to the echo power in a given direction (ϕ,θ). The pattern of $f^2(\phi,\theta)$ typically generates conical or pencial-shaped beams, as shown in Fig. 3.5. It consists of a large main lobe (or main beam) that exhibits the greatest field strength and several smaller lobes surrounding the main lobe, with subsidiary power maxima called side lobes, caused by interference effects. Although side lobes extend outward only a short distance and contain very low power, they can detect strong non-meteorological targets in proximity of the radar, so that strong nearby ground echoes can arise and cause confusion in interpreting close targets. The smaller lobes in directions nearly opposite to the main lobe are called back lobes. The direction of maximum power is the beam axis and the planes $\phi = 0°$ and $\theta = 0°$ are the principal planes of antenna pattern. Another important parameter is the angular width of the main lobe, also called beamwidth,

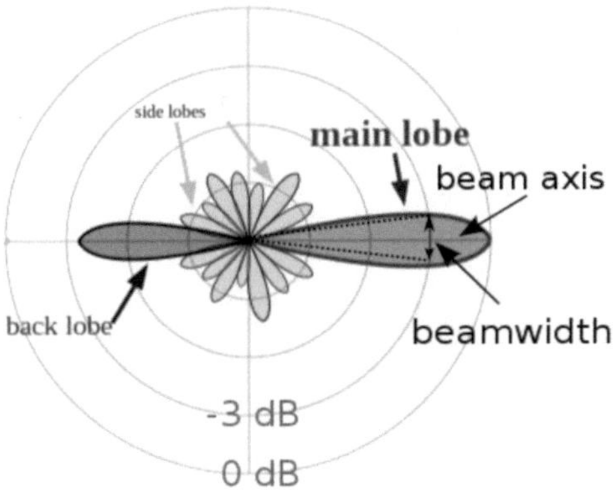

Fig. 3.5.: Antenna radiation pattern: the radial distance from the center represents signal strength.

which is usually defined as the angle between the two directions in a principal plane where the antenna power is one-half, or 3-dB less than its maximum value. The area within these 3-dB points contains nearly 80 percent of all power. Beamwidth varies directly with wavelength and inversely with antenna size. Use of the half-power or 3-dB points to define the beamwidth derive from the so-called “Rayleigh criterion”, according to which two distant points separated by an angle equal to the half-power beamwidth can be resolved. This criterion is not directly applicable to radar because it involves two-way propagation of the microwaves, but the 3-dB beamwidth provides a convenient basis for evaluating and comparing the performance of radar antennas, so it is almost universally employed. Beamwidths are dimensionless and are measured in radians, although for convenience the values are often converted to degrees. The symbols ϕ_3 and θ_3 are used to represent the beamwidths in the horizontal (azimuth) and vertical (elevation) principal planes, respectively (ϕ_3 and θ_3 typically have small values ($\approx 1°$)). For meteorological applications, $f^2(\phi, \theta)$ is usually expressed as a Gaussian function (Probert-Jones, 1962):

$$f^2(\phi, \theta) = \exp\left[-4\ln 2\left(\frac{\phi^2}{\phi_3^2} + \frac{\theta^2}{\theta_3^2}\right)\right] \quad , \tag{3.3}$$

by neglecting the echoes from side lobes because side lobes can be different on different radar systems and a general parametrization formula for the effective beam pattern would be complicated and would not show significant impacts for most practical applications. For symmetric antennas, ϕ_3 and θ_3 are equal, so Eq. (3.3) becomes:

$$f^2(\theta_z) = \exp\left[-4\ln 2\left(\frac{\theta_z^2}{\theta_3^2}\right)\right] \quad , \tag{3.4}$$

with $\theta_z^2 = \phi^2 + \theta^2$.

We have so far neglected that output values in radar data sets are usually averages over many consecutive pulses to achieve statistical signal stability while the antenna rotates. As shown in Blahak (2008a), this leads to a somewhat broader effective beam weighting function, denoted by f_e. Provided that the radar is scanning horizontally at a constant elevation ε_0, which depends on the common practice in setting up radar schedulers scanning azimuthally in a continuous mode with a discrete change in elevation after finishing a 360° rotation, that is (Blahak, 2008a),

$$f_e^2(\alpha,\varepsilon) = \exp\left\{-4\ln 2\left[\left(\frac{(\alpha-\alpha_*)\cos\varepsilon}{\alpha_{3,eff,0} + (\cos\varepsilon_0 - 1)\Delta\alpha[1-\exp(-1.5\Delta\alpha/\theta_3)]}\right)^2 + \left(\frac{\varepsilon-\varepsilon_0}{\theta_3}\right)^2\right]\right\} \quad . \tag{3.5}$$

Note that in order to describe the angular averaging correctly in terms of azimuth, the formulation has been transformed from the beam system to the radar system, i.e, $\vec{r} = (r, \alpha_*, \varepsilon)$, where α_* is the center of the averaging interval. $\Delta\alpha$ is the averaging interval of the consecutive pulses, and $\alpha_{3,eff,0}$ is the effective 3-dB beamwidth at 0° elevation, which depends only on the radar specific ratio $\Delta\alpha/\theta_3$ and can be calculated by interpolation from Table 1 of Blahak (2008a).

A key parameter often used to judge the quality of radar observations is the spatial resolution, defined as the minimum separation between two targets of equal reflectivity that permits them to be distinguished individually in a processed radar image. The spatial resolution at any point in a radar image is determined by computing the resolution in two dimensions: the range and the azimuthal resolutions. The range resolution is the ability of radar to differentiate two targets that are close together in range. The range resolution is limited by the pulse length, the types and sizes of the targets, and the efficiency of the receiver and the indicator, but the pulse length is the primary factor. A well-designed radar system, with all other factors at maximum efficiency, is able to discriminate between separate echoes only if the difference in their delays is larger than the pulse duration τ, so the range resolution Δr is equal to $c\tau/2$. The azimuthal resolution is the ability of radar to resolve between two targets in the azimuthal direction. Two targets can be separated in the azimuthal direction only if the distance between them is larger than one beamwidth, so the beamwidth is taken as the measure of azimuthal resolution. Since radars have a certain spatial resolution, the radar displays usually look gridded and blocky (see Fig. 3.2), and each individual block or box of data is called a pixel, bin or gate.

Radars from DWD's network have range resolution of 1000 m and azimuthal resolution of $1°$. The antenna moves constantly in the azimuthal direction from $0°$ to $359°$ and each pulse is sent in a different azimuthal direction, separated by $1°$. When the radar finishes scanning in 360 degrees at one elevation, it tilts up to the next elevation and does the same sweep again. The radar repeats this until it has scanned at all elevations (see Fig. 3.6). As the beam broadening (see Section 3.6) reduces the reliability of radar measurements with distance, only measurements within the range of 124 km ($= 124 \times 1000$ m) from each radar site are taken into account. As mentioned in Chapter 1, there are different scanning types, i.e., Doppler- and intensity volume scans and precipitation scan. For the purpose of data assimilation, the Doppler volume scan is of interest, therefore the relevant one in this thesis. Currently, one complete Doppler volume scan takes about 5 minutes, so the radars have temporal resolution of 5 minutes.

In Eq. (3.1), W is the range weighting function. To simply notation, we assume that all scatterers located on the same ray path contribute equally to the received power and the range weighting function W can be written as a simple step function:

$$W(r_0 - r) = \begin{cases} 1, & \text{if } r \in [r_0 - \Delta r/2, r_0 + \Delta r/2]; \\ 0, & \text{otherwise.} \end{cases} \tag{3.6}$$

Considering the fact that an output value of the radar results from averaging over several bins in range, Eq. (3.6) also represents a sound approximation. Moreover, for

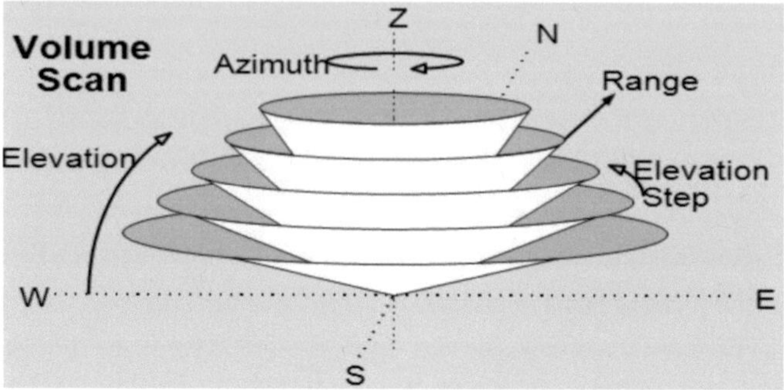

Fig. 3.6.: Sketch of volume scan strategy showing the polar coordinate system (Ruffieux and Illingworth, 2008)

our typical applications the horizontal resolution of the COSMO-model is about one order larger than the pulse length, so that a realistic range weighting would not have significant effects anyway.

It should be remembered that the microwave energy emitted by radar is, in fact, a wave, so it has all the characteristics of waves such as wavelength, defined as the distance between two points of corresponding phase in consecutive cycles and denoted with λ. In the microwave portion of the electromagnetic spectrum, wavelengths vary between 1 mm and 1 m (see Fig. 3.7). According to the wavelength, Doppler radar can be divided into several band categories, which are L, S, C, X, K_u, K and K_a (see Tab. 3.1). The DWD radar network consists of C-band radars.

Band	Wavelength [cm]	Frequency [GHz]	Usage
L	15-30	1-2	clear air turbulence studies
S	8-15	2-4	near and far range weather observation
C	4-8	4-8	short range weather observation
X	2.5-4	8-12	cloud development studies
K_u	1.7-2.5	12-18	satellite communications
K	1.2-1.7	18-27	detecting clouds
K_a	0.8-1.1	27-40	airport surveilance

Tab. 3.1.: Overview on the different bands of Doppler radars

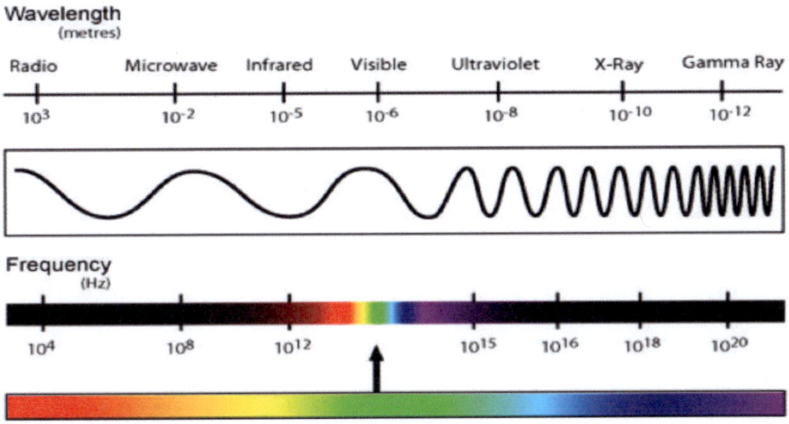

Fig. 3.7.: The electromagnetic spectrum: the microwave region of the spectrum is towards the left, where wavelengths are longer and frequencies are lower.

The typical observations of a conventional Doppler radar are reflectivity and Doppler velocity. Their measurement principles are introduced in the following sections.

3.2. Reflectivity factor

As electromagnetic radiation travels in the atmosphere, it interacts with air molecules, dust particles, water vapor, rain, ice particles, insects and etc.. These interactions make the radiation undergo attenuation in the form of scattering and absorption. The amount, by which a "target" can scatter or absorb radiation, is typically described through an apparent area, called cross section σ.

3.2.1. Cross section

When a target is illuminated by a wave having an incident power density S_i, it will scatter/absorb a part of the power. An observer located at a specific position (ϕ,θ) will receive radiation scattered by the target with a power density S_r. Under the assumption that the target is an isotropic scatterer, σ can be directly calculated by

$$\sigma(r,\phi,\theta) = 4\pi r^2 \frac{S_r(\phi,\theta)}{S_i} \quad , \tag{3.7}$$

where ϕ and θ are referenced to a polar axis connecting the target and the transmitter with the target at the origin, r is the distance between the target and the observer. In general, the scattering cross section depends on the angles ϕ and θ, which means that scattering is not really isotropic. Also note that the value of σ does not correspond to the geometric cross section of the target. Sometimes, σ is also called the differential cross section to distinguish it from the total cross section Q, which is obtained by the integration of σ over the entire solid angle[1]. For instance, the total scattering cross section Q_s multiplied by the power density S_i is equivalent to total amount of energy removed from the electromagnetic wave due to scatter in all directions. A certain amount of energy is absorbed and heats the target. The amount of energy removed from the electromagnetic wave through this process is equal to the total absorption cross section Q_a multiplied by S_i. The cumulative effect of scattering and absorption is described by the attenuation cross section Q_t. For the radar technique, the value of $\sigma(\phi, \theta)$ in the direction from which the wave originates is of great interest, this value defines the backscattering cross section σ_b.

On the basis of application of Maxwell's equations to the scattering of a planar wave by a homogeneous sphere in a nonabsorbing medium, Mie (1908) formulated a complete scattering/absorption theory in terms of an infinite series of electric and magnetic multipoles. The attenuation, total scattering, total absorption and backscattering cross sections can be expressed as

$$Q_t = \frac{\lambda^2}{2\pi}(-Re)\left[\sum_{n=1}^{\infty}(2n+1)(a_n+b_n)\right] \quad , \tag{3.8}$$

$$Q_s = \frac{\lambda^2}{2\pi}\sum_{n=1}^{\infty}(2n+1)(|a_n|^2+|b_n|^2) \quad , \tag{3.9}$$

$$Q_a = Q_t - Q_s \quad , \tag{3.10}$$

$$\sigma_b = \frac{\lambda^2}{4\pi}\left|\sum_{n=1}^{\infty}(-1)^n(2n+1)(a_n-b_n)\right|^2 \quad , \tag{3.11}$$

respectively, where a_n and b_n are the so-called Bessel functions, which represent the magnetic and electric multipoles of order n and depend on the radio electric size $\gamma = \pi D/\lambda$ and the complex refractive index m.

[1]In some literatures, Q will normalized by 4π, in that case Q will be actually an averaged cross section.

For relatively small drop-diameter/wavelength ratios (i.e., $D \leq \lambda/16$), the multipole moments can be neglected and only the lowest order in the series solution, namely the dipolterm is considered, which is equivalent to setting all a_n and b_n to zero except a_1,

$$a_1 = \frac{2i}{3}\left(\frac{\mathrm{m}^2-1}{\mathrm{m}^2+2}\right)\gamma^3 \quad , \tag{3.12}$$

which yields

$$Q_s = \frac{2\,\pi^2}{3\,\lambda^4}\,|K|^2 D^6 \quad , \tag{3.13}$$

$$Q_a = \frac{\pi^2}{\lambda} Im(-K) D^3 \quad , \tag{3.14}$$

$$\sigma_b = \frac{\pi^5}{\lambda^4}\,|K|^2 D^6 \quad , \tag{3.15}$$

where $K = \left|\frac{\mathrm{m}^2-1}{\mathrm{m}^2+2}\right|^2$ is the dielectric factor of the particles. Eqs. (3.13)-(3.15) are the Rayleigh approximation, according to which σ_b is proportional to the reciprocal of the fourth power of the wavelength and to the sixth power of the particle's diameter.

3.2.2. Unattenuated reflectivity

The summation of all backscattering cross sections per unit volume is defined as the radar reflectivity and represented by the symbol η:

$$\eta = \sum_{k\in S}\sum_{j\in N^k} \sigma_{b_{kj}} \quad , \tag{3.16}$$

where S contains all hydrometeor types in volume and N^k the number of hydrometeors of type k, or in continuous form:

$$\eta = \sum_{k\in S}\int_0^\infty \sigma_{b_k}(D)\mathcal{N}^k(D)dD \quad , \tag{3.17}$$

where $\mathcal{N}^k$ [$\mathrm{mm^{-1}m^{-3}}$] is the particle size distribution (PSD) and $\mathcal{N}^k(D)dD$ represents the mean number of hydrometeors with equivalent spherical diameters between D and $D+dD$ [mm] present per unit volume.

At standard wavelengths of the weather radars, the conditions for Rayleigh scattering in terms of clouds and precipitation (except hail) are usually fulfilled, so we have

$$\eta = \sum_{k \in S} \frac{\pi^5}{\lambda^4} |K_k|^2 \underbrace{\int_0^\infty \mathcal{N}^k(D_k) D_k^6 dD_k}_{:=Z_k} = \sum_{k \in S} \frac{\pi^5}{\lambda^4} |K_k|^2 Z_k \quad . \tag{3.18}$$

As indicated by the factor λ^{-4} in Eq. (3.18), η varies strongly with wavelength, which makes η not suitable for describing precipitation targets when weather radars of different wavelengths are involved. The new variable Z_k , under Rayleigh approximation, is the so-called radar reflectivity factor, which depends exclusively on the number of scattering elements and their sizes. In other words, the reflectivity factor Z_k is a typical feature of the target.

In practice, the phase and composition of hydrometeors inside the volume is usually unknown. Unter the assumption that hydrometeors are spherical liquid hydrometeors and satisfies Rayleigh approximation, it yields

$$\eta = \frac{\pi^5}{\lambda^4} |K_w|^2 Z_e \quad , \tag{3.19}$$

where $|K_w|^2 = 0.93$ is the dielectric factor for water and Z_e is the (water-)equivalent reflectivity factor that is the most important variable in radar meteorology.

Z_e has conventional units of $[\,\mathrm{mm}^6/\mathrm{m}^3]$. Because numerical values of Z_e may span several orders of magnitude in practice, it is convenient for numerical calculation to use a logarithmic scale. The logarithmic equivalent reflectivity factor is defined as $10\log_{10}\left[\frac{Z_e}{1\,\mathrm{mm}^6/\mathrm{m}^3}\right]$ and is expressed in units of dBZ (Battan, 1973). For instance, when $Z_e = 1\,\mathrm{mm}^6/\mathrm{m}^3$ the reflectivity factor level is 0 dBZ and when $Z_e = 10^5\,\mathrm{mm}^6/\mathrm{m}^3$

it is 50 dBZ. It is worth noting that the difference of two logarithmic reflectivities is not in logarithmic units but in linear units of dB:

$$\text{dBZ} - \text{dBZ} \longrightarrow \text{dB} \quad . \tag{3.20}$$

For example, a 55 dBZ storm has decreased 10 dB to become a 45 dBZ storm, or say, the intensity difference between a 55 dBZ storm and a 45 dBZ storm is 10 dB.

3.2.3. Attenuation

It can be shown that $Q_s < Q_a$ for the Rayleigh approximation, so one would use Q_a for attenuation estimation at wavelengths $\lambda \geq 10$ cm, at which the Rayleigh condition $D \leq \lambda/16$ is fulfilled for all raindrops. However, we must note that the Rayleigh approximation is just the leading term in the series solution formulated by Mie but the other terms of the series contributes significantly to absorption (even at $\lambda = 10$ cm although $D \leq \lambda/16$ for all raindrops). Consequently, the Rayleigh approximation for attenuation is in error for moderate to heavy rains and we must employ the Mie solution for Q_t.

The attenuation for a given path in the atmosphere is described by the Beer-Lambert law. At a particular range r, the attenuation factor of the radar beam is expressed by

$$\ell(r) = \exp\left(\int_0^r \Lambda(r')dr'\right) \tag{3.21}$$

is the one-way loss factor due to attenuation, where

$$\Lambda(r) := \sum_{k\in S} \int_0^\infty Q_{t_k}(D,r)\mathcal{N}^k(D,r)dD \tag{3.22}$$

is attenuation coefficient.

Another often used term is the so-called two-way attenuation coefficient k_2, which is defined as

$$k_2 = 10\log_{10}(\exp(-2\Lambda)) = -\frac{20\Lambda}{\ln 10} \tag{3.23}$$

in a logarithmic scale [dB km^{-1}], where the power -2 is the two-way factor.

The final form of the attenuated reflectivity for spherical particles at a particular distance r and wavelength λ is given by

$$Z_e = \frac{\eta \lambda^4}{\pi^5 |K_w|^2} \ell^{-2}(r) \tag{3.24}$$

with the power -2 for the two-way attenuation.

Loosely speaking, attenuation increases as radar wavelength decreases, for instance, radars operated with $\lambda = 5\,\text{cm}$ suffer 100 times larger power loss than radars with $\lambda = 10\,\text{cm}$, therefore attenuation could have considerable influences on observations from the C-Band radar network of DWD. For instance, Scarchilli et al. (1993) stated that specific attenuation could attain $0.5\,\text{dB/km}$ for C-band radars. As known, the attenuation of the radar signal arises from absorption and scattering by atmospheric gases and hydrometeors, but in fact the attenuation by gases is often a small constant and already corrected in most radar signal processors, thus it is neglected in this work.

3.2.4. Radar equation

The radar equation provides the fundamental relationship between the received power and the characteristics of the target, situated at position $\vec{r}_0 = (r_0, \alpha_0, \varepsilon_0)$, as a function of the technical characteristics of the radar and the atmospheric conditions on the propagation path.

As shown in Doviak and Zrnic (1993), under assumptions that:

1. the particles occupy the entire volume of the pulse,

2. the hydrometeor particles are homogeneous dielectric spheres with diameters small compared to the radar wavelength,

3. all the particles have the same dielectric factor $|K|^2$,

4. the main lobe of the antenna beam pattern is expressed by a Gaussian function f,

5. the incident and back-scattered waves are linearly polarized,

6. radar miscalibration and wetted radome are negligible,

the radar equation can be written in terms of the radar reflectivity (Blahak, 2004):

$$P_r(\vec{r_0}) = C_r \int_{r_0-\Delta r/2}^{r_0+\Delta r/2} \int_{-\pi}^{\pi} \int_{-\pi/2}^{\pi/2} \eta(r,\phi,\theta)\ell^{-2}(r,\phi,\theta)\frac{f^4(\phi,\theta)}{r^4}|W(r_0-r)|^2 r^2 \cos\theta d\theta d\phi dr \quad , \tag{3.25}$$

where P_r is the received power at the antenna. As shown in Eq. (3.25), η is integrated over pulse volume ($dV = r^2 \cos\theta d\theta d\phi dr$), weighted by the antenna pattern $f^2(\theta,\phi)$ and the range weighting function W.

By substituting Eq. (3.6) in Eq. (3.25) we obtain

$$P_r(\vec{r_0}) = C_r \int_{r_0-\Delta r/2}^{r_0+\Delta r/2} \int_{-\pi}^{\pi} \int_{-\pi/2}^{\pi/2} \eta(r,\phi,\theta)\ell^{-2}(r,\phi,\theta)\frac{f^4(\phi,\theta)}{r^2} \cos\theta d\theta d\phi dr \quad . \tag{3.26}$$

Note again that the simple boxcar-function for the range weighting is a good approximation if several range bins are averaged in range by the radar processor to gain a single output value.

Applying the mean value theorem to Eq. (3.26) yields:

$$P_r(\vec{r_0}) = C_r \overline{\left[\frac{\eta}{\ell^2}\right]}(\vec{r_0}) \int_{r_0-\Delta r/2}^{r_0+\Delta r/2} \int_{-\pi}^{\pi} \int_{-\pi/2}^{\pi/2} \frac{f^4(\phi,\theta)}{r^2} \cos\theta d\theta d\phi dr \quad , \tag{3.27}$$

where

$$\overline{\left[\frac{\eta}{\ell^2}\right]}(\vec{r_0}) = \frac{\int_{r_0-\Delta r/2}^{r_0+\Delta r/2} \int_{-\pi}^{\pi} \int_{-\pi/2}^{\pi/2} \eta(r,\phi,\theta)\ell^{-2}(r,\phi,\theta)\frac{f^4(\phi,\theta)}{r^2} \cos\theta d\theta d\phi dr}{\int_{r_0-\Delta r/2}^{r_0+\Delta r/2} \int_{-\pi}^{\pi} \int_{-\pi/2}^{\pi/2} \frac{f^4(\phi,\theta)}{r^2} \cos\theta d\theta d\phi dr} \quad . \tag{3.28}$$

Since attenuation by hydrometeors is not related to the beam weighting function, we can set $\ell^2 = 1$ and obtain

$$P_r(\vec{r}_0) = C_r \overline{\eta}(\vec{r}_0) \int\limits_{r_0-\Delta r/2}^{r_0+\Delta r/2} \int\limits_{-\pi}^{\pi} \int\limits_{-\pi/2}^{\pi/2} \frac{f^4(\phi,\theta)}{r^2} \cos\theta d\theta d\phi dr \quad , \tag{3.29}$$

or equivalently,

$$\overline{\eta}(\vec{r}_0) = \frac{P_r(\vec{r}_0)}{C_r} \frac{1}{\int\limits_{r_0-\Delta r/2}^{r_0+\Delta r/2} \int\limits_{-\pi}^{\pi} \int\limits_{-\pi/2}^{\pi/2} \frac{f^4(\phi,\theta)}{r^2} \cos\theta d\theta d\phi dr} \quad . \tag{3.30}$$

To relate the received power to the physical properties of the medium observed, η is substituted by the factor Z using Eq. (3.19), which results in

$$\overline{Z}(\vec{r}_0) = \frac{\lambda^4}{\pi^5 |K|^2} \frac{P_r(\vec{r}_0)}{C_r} \frac{1}{\int\limits_{r_0-\Delta r/2}^{r_0+\Delta r/2} \int\limits_{-\pi}^{\pi} \int\limits_{-\pi/2}^{\pi/2} \frac{f^4(\phi,\theta)}{r^2} \cos\theta d\theta d\phi dr} \quad . \tag{3.31}$$

However, in day-to-day radar operations there are many occasions where one or more of these assumed conditions are violated. For instance, Eq. (3.26) does not hold if the raindrops and ice particles illuminated by the radar beam are not in the Rayleigh regime. If conditions 2 and 3 are not satisfied[2], Z is replaced by Z_e, which yields

$$P_r(\vec{r}_0) =$$

$$C_r \frac{\pi^5 |K_w|^2}{\lambda^4} \int\limits_{r_0-\Delta r/2}^{r_0+\Delta r/2} \int\limits_{-\pi}^{\pi} \int\limits_{-\pi/2}^{\pi/2} Z_e(r,\phi,\theta) \ell^{-2}(r,\phi,\theta) \frac{f^4(\phi,\theta)}{r^2} \cos\theta d\theta d\phi dr \quad . \tag{3.32}$$

Let now Z_e be the "true" effective radar reflectivity field, $Z_e^{(R)}$ an instantaneous value derived by the radar processor from an instantaneous power measurement $P_r^{(R)}$. As the radar constant C_r is generally not exactly known and may change with time, we replace it in Eq. (3.31) by a reference radar constant C_0, which is used in radar software.

[2]Considering the violation of the other conditions are beyond the scope of my work.

By using Eq. (3.32) and substituting $\overline{Z}$ with $Z_e^{(R)}$ and P_r with $P_r^{(R)}$, Eq. (3.31) can be written as

$$Z_e^{(R)}(\vec{r}_0) = \frac{\lambda^4}{\pi^5|K_w|^2} \frac{P_r^{(R)}(\vec{r}_0)}{C_r} \frac{1}{\int\limits_{r_0-\Delta r/2}^{r_0+\Delta r/2} \int\limits_{-\pi}^{\pi} \int\limits_{-\pi/2}^{\pi/2} \frac{f^4(\phi,\theta)}{r^2} \cos\theta d\theta d\phi dr} \tag{3.33}$$

$$= \frac{\int\limits_{r_0-\Delta r/2}^{r_0+\Delta r/2} \int\limits_{-\pi}^{\pi} \int\limits_{-\pi/2}^{\pi/2} Z_e(r,\phi,\theta)\ell^{-2}(r,\phi,\theta)\frac{f^4(\phi,\theta)}{r^2} \cos\theta d\theta d\phi dr}{\int\limits_{r_0-\Delta r/2}^{r_0+\Delta r/2} \int\limits_{-\pi}^{\pi} \int\limits_{-\pi/2}^{\pi/2} \frac{f^4(\phi,\theta)}{r^2} \cos\theta d\theta d\phi dr} . \tag{3.34}$$

Eqs. (3.33-3.34) establish the relationship between the single measured $Z_e^{(R)}$ and received power $P_r^{(R)}$. Eq. (3.33) is applied in most of the radar software while Eq. (3.34) is for our simulation of interest. However, P_r can not be estimated just by a single radar pulse. Marshall and Hitschfeld (1953) showed that owing to the random distribution of the scatterers relative to the pulse wave phase, the squared amplitude of the electron magnetic field derived from a single $P_r^{(R)}$ is statistically distributed in the vicinity of P_r and only its ensemble average $\langle P_r^{(R)} \rangle$ is equal to P_r. The common technique to achieve $\langle P_r^{(R)} \rangle$ with statistical stability is averaging over many consecutive pulses during the antenna rotation operating with a certain angular velocity. As mentioned in Section 3.6, this leads to a somewhat broader effective beam weighting function f_e as expressed in Eq. 3.5. Substituting f with f_e in Eq. (3.33) yields

$$\langle Z_e^{(R)} \rangle(\vec{r}_0) = \frac{C_r}{C_0} \frac{\int\limits_{r_0-\Delta r/2}^{r_0+\Delta r/2} \int\limits_{\alpha_*-\pi}^{\alpha_*+\pi} \int\limits_{\varepsilon_0-\pi/2}^{\varepsilon_0+\pi/2} Z_e(r,\alpha,\varepsilon)\ell^{-2}(r,\alpha,\varepsilon)\frac{f_e^4(\alpha,\varepsilon)}{r^2} \cos\varepsilon d\varepsilon d\alpha dr}{\int\limits_{r_0-\Delta r/2}^{r_0+\Delta r/2} \int\limits_{\alpha_*-\pi}^{\alpha_*+\pi} \int\limits_{\varepsilon_0-\pi/2}^{\varepsilon_0+\pi/2} \frac{f_e^4(\alpha,\varepsilon)}{r^2} \cos\varepsilon d\varepsilon d\alpha dr} , \tag{3.35}$$

which takes azimuthal scanning into account.

On the basis of Eq. (3.35), it is often assumed that the distance from the radar antenna to the target is large compared to the length of pulse volume (i.e., $r_0 \gg \Delta r/2$)

and that Z_e and ℓ do not vary much within Δr, the integrand $1/r^2 \approx 1/r_0^2$ can thus be pulled out of the integral, which results in:

$$\langle Z_e^{(R)} \rangle (\vec{r}_0) = \frac{C_r}{C_0} \frac{\int\limits_{\alpha_*-\pi}^{\alpha_*+\pi} \int\limits_{\varepsilon_0-\pi/2}^{\varepsilon_0+\pi/2} Z_e(r_0, \alpha, \varepsilon) \ell^{-2}(r, \alpha, \varepsilon) f_e^4(\alpha, \varepsilon) \cos\varepsilon d\varepsilon d\alpha}{\int\limits_{\alpha_*-\pi}^{\alpha_*+\pi} \int\limits_{\varepsilon_0-\pi/2}^{\varepsilon_0+\pi/2} f_e^4(\alpha, \varepsilon) \cos\varepsilon d\varepsilon d\alpha} \quad . \tag{3.36}$$

Since the factor C_r/C_0 is generally unknown, it is set to 1 in the remainder of this thesis with caution that it may pose a serious source of bias (cf. Section 3.4).

For the purpose of brevity, Z_e is often called reflectivity factor or simply reflectivity. To avoid terminology ambiguity, the term reflectivity will hereafter always refer to the effective reflectivity factor Z_e.

3.3. Doppler velocity

A Doppler radar observes not just reflectivity but also mean Doppler velocity within the resolution volume (function of the mean component of scatterers' three-dimensional motion in the radial direction toward or away from the radar), since each scatter has its own speed and direction, depending on its size, shape and motion of the surrounding air.

The measurement of Doppler velocity is based on the phenomenon called the Doppler effect. When a pulse of radiation interacts with a target, it induces molecular vibrations of the target's electric and magnetic fields. If the target is moving toward the transmitter in the radial direction at velocity v_r (positive being away from the radar), its vibrational frequency gets higher by v_r/λ (Doviak and Zrnic, 1993). The vibrating molecules themselves create electromagnetic fields, which in turn radiate outward from the target. The frequency of scattered radiation is then shifted by an amount of $\Delta f = -2v_r/\lambda$. The factor of 2 is due to a two-step increase in the frequency, first in the target's electric vibrational frequency and second in the frequency of its radiation field in the radial direction.

The maximum velocity $v_{r,max}$ that a Doppler radar can detect unambiguously is given by the velocity, which just produces a phase shift of $\pm\pi$. This velocity is called Nyquist velocity, given by (Battan, 1973)

$$|v_{r,max}| = \text{PRF}\ \lambda/4 \quad . \tag{3.37}$$

Notice that the maximum unambiguous range r_{max} and maximum unambiguous velocity $v_{r,max}$ both depend on PRF but in opposite ways, which leads to the fundamental equation:

$$|v_{r,max} \cdot r_{max}| = c\lambda/8 \quad . \tag{3.38}$$

This is known as "Doppler dilemma", a trade-off has to be made between $v_{r,max}$ and r_{max}. For a typical C-band weather radar with $r_{max} = 150\,\text{km}$, $v_{r,max}$ is about only $12\,\text{m/s}$. When an environmental wind exceeds $v_{r,max}$, the radar interprets it as a weaker wind of the opposite sign. The true environmental wind is offset by factor of $2 \cdot v_{r,max}$ until it falls within the Nyquist interval. This is called velocity folding/aliasing. For example, if $v_{r,max}$ is $25\,\text{m/s}$ and the environmental wind speed is $-30\,\text{m/s}$, then it is folded and the radar interprets it as $20\,\text{m/s}$. Nowadays, advanced techniques like dual-PRF (Dazhang et al., 1984) and correction by using dealiasing algorithms (Haase and Landelius, 2004) can be used to mitigate the ambiguity problem.

Doppler velocity is also an observed volumetric quantity. Unlike the pulse volume averaged reflectivity, this is the average of point velocities weighted by the reflectivity and antenna pattern in the pulse volume. The relationship between the point velocities, reflectivity fields, antenna pattern and the power weighted moment is given in Doviak and Zrnic (1993). Here we refine the original formulation by taking the fall velocity of hydrometeors and attenuation into account and the averaged Doppler velocity is given by

$$
\begin{aligned}
&\langle v_r^{(R)}\rangle(\vec{r}_0) = \\
&\frac{\int\limits_{r_0-\Delta r/2}^{r_0+\Delta r/2}\int\limits_{-\pi}^{+\pi}\int\limits_{-\pi/2}^{\pi/2} \ell^{-2}(r,\phi,\theta)\left(\int\limits_0^{\infty}\sigma_b(D,r,\phi,\theta)\mathscr{N}(D,r,\phi,\theta)[(\vec{v}(r,\phi,\theta)-w_t(D)\vec{e}_3)\cdot\vec{e}_r]dD\right)\frac{f^4(\phi,\theta)}{r^2}\cos\theta d\theta d\phi dr}{\int\limits_{r_0-\Delta r/2}^{r_0+\Delta r/2}\int\limits_{-\pi}^{+\pi}\int\limits_{-\pi/2}^{\pi/2}\frac{\eta(r,\phi,\theta)}{\ell^2(r,\phi,\theta)}\frac{f^4(\phi,\theta)}{r^2}\cos\theta d\theta d\phi dr} \\
&= \frac{\int\limits_{r_0-\Delta r/2}^{r_0+\Delta r/2}\int\limits_{-\pi}^{+\pi}\int\limits_{-\pi/2}^{\pi/2} \ell^{-2}(r,\phi,\theta)\left(\int\limits_0^{\infty}\sigma_b(D,r,\phi,\theta)\mathscr{N}(D,r,\phi,\theta)dD\right)\vec{v}(r,\phi,\theta)\vec{e}_r\frac{f^4(\phi,\theta)}{r^2}\cos\theta d\theta d\phi dr}{\int\limits_{r_0-\Delta r/2}^{r_0+\Delta r/2}\int\limits_{-\pi}^{+\pi}\int\limits_{-\pi/2}^{\pi/2}\frac{\eta(r,\phi,\theta)}{\ell^2(r,\phi,\theta)}\frac{f^4(\phi,\theta)}{r^2}\cos\theta d\theta d\phi dr} \\
&\quad - \frac{\int\limits_{r_0-\Delta r/2}^{r_0+\Delta r/2}\int\limits_{-\pi}^{+\pi}\int\limits_{-\pi/2}^{\pi/2} \ell^{-2}(r,\phi,\theta)\left(\int\limits_0^{\infty}\sigma_b(D,r,\phi,\theta)\mathscr{N}(D,r,\phi,\theta)w_t(D)dD\right)(\vec{e}_3\cdot\vec{e}_r)\frac{f^4(\phi,\theta)}{r^2}\cos\theta d\theta d\phi dr}{\int\limits_{r_0-\Delta r/2}^{r_0+\Delta r/2}\int\limits_{-\pi}^{+\pi}\int\limits_{-\pi/2}^{\pi/2}\frac{\eta(r,\phi,\theta)}{\ell^2(r,\phi,\theta)}\frac{f^4(\phi,\theta)}{r^2}\cos\theta d\theta d\phi dr} \\
&= \frac{\int\limits_{r_0-\Delta r/2}^{r_0+\Delta r/2}\int\limits_{-\pi}^{+\pi}\int\limits_{-\pi/2}^{\pi/2} (\vec{v}(r,\phi,\theta)\cdot\vec{e}_r)\frac{\eta(r,\phi,\theta)}{\ell^2(r,\phi,\theta)}\frac{f^4(\phi,\theta)}{r^2}\cos\theta d\theta d\phi dr}{\int\limits_{r_0-\Delta r/2}^{r_0+\Delta r/2}\int\limits_{-\pi}^{+\pi}\int\limits_{-\pi/2}^{\pi/2}\frac{\eta(r,\phi,\theta)}{\ell^2(r,\phi,\theta)}\frac{f^4(\phi,\theta)}{r^2}\cos\theta d\theta d\phi dr} \\
&\quad - \frac{\int\limits_{r_0-\Delta r/2}^{r_0+\Delta r/2}\int\limits_{-\pi}^{+\pi}\int\limits_{-\pi/2}^{\pi/2} (\vec{e}_3\cdot\vec{e}_r)\overline{w}_t(r,\phi,\theta)\frac{\eta(r,\phi,\theta)}{\ell^2(r,\phi,\theta)}\frac{f^4(\phi,\theta)}{r^2}\cos\theta d\theta d\phi dr}{\int\limits_{r_0-\Delta r/2}^{r_0+\Delta r/2}\int\limits_{-\pi}^{+\pi}\int\limits_{-\pi/2}^{\pi/2}\frac{\eta(r,\phi,\theta)}{\ell^2(r,\phi,\theta)}\frac{f^4(\phi,\theta)}{r^2}\cos\theta d\theta d\phi dr} , \qquad (3.39)
\end{aligned}
$$

where $\vec{v} = \begin{pmatrix} u \\ v \\ w \end{pmatrix}$ is the 3D wind vector, $\vec{e_3} = \begin{pmatrix} 0 \\ 0 \\ 1 \end{pmatrix}$ is the unit vector upwards perpendicular to the earth surface and $\vec{e_r} = \begin{pmatrix} \cos\theta\sin\phi \\ \cos\theta\cos\phi \\ \sin\theta \end{pmatrix}$ is the unit vector on the radial ray path direction.

In analogy with reflectivity, by neglecting the integral over r and applying the effective beam weighting function f_e, we obtain

$$\langle v_r^{(R)}\rangle(\vec{r_0}) = \frac{\int\limits_{\alpha_*-\pi}^{\alpha_*+\pi}\int\limits_{\varepsilon_0-\pi/2}^{\varepsilon_0+\pi/2} (\vec{v}(r_0,\alpha,\varepsilon)\cdot\vec{e_r})\frac{\eta(r_0,\alpha,\varepsilon)}{\ell^2(r_0,\alpha,\varepsilon)}f_e^4(\alpha,\varepsilon)\cos\varepsilon d\varepsilon d\alpha}{\int\limits_{\alpha_*-\pi}^{\alpha_*+\pi}\int\limits_{\varepsilon_0-\pi/2}^{\varepsilon_0+\pi/2}\frac{\eta(r_0,\alpha,\varepsilon)}{\ell^2(r_0,\alpha,\varepsilon)}f_e^4(\alpha,\varepsilon)\cos\varepsilon d\varepsilon d\alpha} - \frac{\int\limits_{\alpha_*-\pi}^{\alpha_*+\pi}\int\limits_{\varepsilon_0-\pi/2}^{\varepsilon_0+\pi/2} (\vec{e_3}\cdot\vec{e_r})\overline{w}_t(r_0,\alpha,\varepsilon)\frac{\eta(r_0,\alpha,\varepsilon)}{\ell^2(r_0,\alpha,\varepsilon)}f_e^4(\alpha,\varepsilon)\cos\varepsilon d\varepsilon d\alpha}{\int\limits_{\alpha_*-\pi}^{\alpha_*+\pi}\int\limits_{\varepsilon_0-\pi/2}^{\varepsilon_0+\pi/2}\frac{\eta(r_0,\alpha,\varepsilon)}{\ell^2(r_0,\alpha,\varepsilon)}f_e^4(\alpha,\varepsilon)\cos\varepsilon d\varepsilon d\alpha}. \quad (3.40)$$

Eq. (3.40) is the benchmark formulation for Doppler velocity in this work.

3.4. Sources of errors

Ideally, we would like to measure with radars the exact local values (i.e. point measurement) of unattenuated Z_e and dealiased v_r in the atmosphere, which is by all means impossible in reality. The accuracy of weather radar measurements varies considerably with radar range, radar types, storm characteristics, geographical location and data processing techniques (Wilson and Brandes, 1979; Dalezios and Kouwen, 1982). Various sources of errors which have been discussed by several authors (Austin, 1987; Joss and Waldvogel, 1990; Wilson and Brandes, 1979; Zawadzki, 1973) are:

1. Non-meteorological echoes (e.g., group clutter and variable clutter);

2. Side lobe echoes;

3. Multiple scattering;

4. Second trip echo: a radar assumes that any returned echo is from the most recent pulse that it has transmitted. If the first pulse reaches clouds farther than r_{max}, before its echoes return, a second pulse has been emitted. When the echoes from the first pulse return, the radar think they are from the second pulse and accordingly places them closer to radar than where they actually are;

5. Aliasing;

6. Beam shielding by natural obstacles (such as mountains and trees) or by man-made obstacles (such as buildings and power poles). At ranges beyond, reflectivities will be undervalued and Doppler velocities will be estimated from a higher altitude than expected;

7. Deviation of atmospheric conditions from assumption of 4/3 Earth radius model (see Section 3.5), which can easily make an error of 100 m in height estimate at far range. For a vertical wind shear of 4 m/s per km, a height bias of 100 is sufficient to produce a 0.4 m/s wind bias;

8. Non-uniform beam filling;

9. Attenuation of radar signals due to heavy rainfall along the beam and the effects of water on the radome;

10. Instrumental noise;

11. Statistical fluctuations of the reflectivity due to the random phase position of the instantaneously received signal, even in case of averaging over several pulses;

12. Radar miscalibration due to unknown drifts in radar constant C_r;

13. Inaccuracies in position specification of azimuth, elevation or radial distance.

On the other hand, a success of data assimilation system generally requires:

(1) Observations are free of bias,

(2) Oberservation error variances and correlations must be properly specified and easy to use,

(3) The operator should simulate observations as accurately as possible.

With respect to radar data, data assimilation is even more challenging because radar data have very high spatial and temporal resolutions than the other observations (see Chapter 1) and there is very little that can invalidate radar data if they are wrong or wrongly assimilated. Consequently, they must be particularly carefully treated in assimilation. Regarding item (1), a strict quality control of radar data being available at real-time is mandatory. In DWD, the quality control consists of two steps: pre- and postprocesses (Hengstebeck et al., 2010). The preprocessed data are also called basic data. In the preprocessing procedure that is done at the radar site itself (within the radar device's signal processor) a first quality control is performed by setting filters and thresholds (e.g., Doppler Filter for removing stationary clutter (Seltmann, 2000)). The basic data from the radar network gathered by an automated file distribution system in real-time at the DWD central office in Offenbach and are ready for the post-processing quality control, by which the remaining errors (e.g., sources of errors 1-5 in the list above) are specified and identified by flags in a quality product. The quality products are made for all radar basic data and archived together with the data themselves in a database system using the BUFR format, which is a binary universal form of representation of meteorological data (Dragosavac, 2008). Item (2) is crucial for the quality of analysis because error variances affect to what extent the forecast fields will be corrected to match observations, and error correlations provide how the observed information will be smoothed in the model space if there is a mismatch between the model resolution and the density of observations. Observation error variances are mainly specified according to the knowledge of instrumental characteristics, which can be estimated by using collocated observations. Observation error correlations are difficult to estimate, so most models of covariances used in practice often assume them to be zero, i.e., one believes that distinct measurements are affected by physically independent errors. This might be reasonable for pairs of observations carried out by distinct instruments but is likely not valid for sets of observations performed by the same platform, like radiosonde, satellite or radar measurements. Neglection of observation error correlations will overrate the weight given to the observations and can create problems in the numerics of the analysis. Usually, considerable error correlations occur when observations are close to each other, so it makes sense to try to minimize them by thinning dense data (see Section 6.8). Concerning item (3), this statement is the main goal of this work. A radar operator uses model variables to simulate expected values of radar measurements. If a data assimilation system can precisely mimic what a radar would observe under given known atmospheric conditions, then it can ingest the true radar data to tune the model states until simulated observations converge towards the true ones. Vice versa, if the simulated observations are biased, wrong information will be assimilated. To minimize the bias, i.e., the difference between observations and simulated observations, we aim to develop an accurate operator that can reproduce radar observations as well as possible by taking each physical process in radar measurements into account, including the sources of errors 6-9.

In the next sections, we will give a detailed description about how the individual physical processes, such as beam bending and broadening, and radar observables (i.e., reflectivity and Doppler velocity) are simulated within the operator, followed by the corresponding sensitive experiments.

3.5. Beam bending

A radar beam which is propagated through the atmosphere encounters variations of refractive index along its trajectory, which causes the beam to become curved. The total angular refraction of the beam between two points is commonly called "bending".

It is helpful to briefly recall the physical basis for computation of atmospheric refraction at first. More thorough treatments on the subject may be found, e.g., in Bean and Dutton (1966) as well as Doviak and Zrnic (1993).

For describing the ray path, the classical geometric optics is a commonly used approximation. This approximation is applicable, if, within one wavelength of radiation,

- the refractive index n changes only very little, and
- the mutual distances between "neighboring" rays change also only very little.

Under these conditions, a single ray path is determined by Fermat's principle, which states that the travel time t between two points A and B be minimal. Travel time depends on the propagation speed, which is given by $c' = c/\mathsf{n}$, where c is the speed of light in vacuum and n is the refractive index of the medium (to be precise: its real part).

Fermat's principle reads

$$t = \int_A^B \frac{\mathsf{n}}{c} dr \stackrel{!}{=} \min \quad , \tag{3.41}$$

where dr is an infinitesimal arc element. This is a classical problem of functional analysis. For the atmospheric ray propagation, it is assumed that the earth is spherical with radius R_E (defining the mean sea level MSL). The refractive index n, in general, varies in all spatial directions, but in the atmosphere, vertical variations are usually much larger than horizontal variations. Therefore, it is assumed hereafter that n only depends on height h over the Earth's surface, which allows us to define the ray path by the Euler-Lagrange-equation of the system, which reads, after transforming the infinitesimal line element dr to the arc distance element ds at MSL height $h = 0$ (Hartree et al., 1946)

$$\frac{d^2h}{ds^2} - \left(\frac{dh}{ds}\right)^2 \left(\frac{1}{\mathsf{n}}\frac{d\mathsf{n}}{dh} + \frac{2}{R_E+h}\right) - \left(\frac{R_E+h}{R_E}\right)^2 \left(\frac{1}{\mathsf{n}}\frac{d\mathsf{n}}{dh} + \frac{1}{R_E+h}\right) = 0 \quad . \tag{3.42}$$

One can show that this second order non-linear ordinary differential equation (ODE) is "almost" equivalent to the integral conserved quantity

$$\mathsf{n}(h)\,(R_E+h)\cos\varepsilon \;=\; const \quad , \tag{3.43}$$

where ε is the local elevation and is given by

$$\tan\varepsilon \;=\; \frac{R_E}{R_E+h}\frac{dh}{ds} \quad . \tag{3.44}$$

By "almost" we mean that integration is mathematically not an equivalent transformation and additional (non-physical) solutions can be created by integration. Here, the is manifested by a sign ambiguity of ε in Eq. (3.43), because $\cos\varepsilon = \cos(-\varepsilon)$. We will come to this problem later in Subsection 3.5.3.

Eq. (3.43) is the well-known Snell's law for a continuous spherically stratified medium, which states that the constant on the r.h.s is conserved along a ray path. This conserved quantity has a similar significance as, e.g., the mechanical energy for the equation of motion.

A useful simplified approximation of Eq. (3.42) may be obtained by assuming:

- $R_E + h \approx R_E$
- $\frac{dh}{ds} \ll 1$ (rays at low elevations)
- $\mathsf{n} \approx 1$

so that Eq. (3.42) becomes

$$\frac{d^2h}{ds^2} = \frac{1}{R_E} + \frac{d\mathsf{n}}{dh} \quad . \tag{3.45}$$

Based on Eq. (3.45), Doviak and Zrnic (1993) showed that the curvature of the ray C_0 is

$$\mathsf{C}_0 = -\frac{d\mathsf{n}}{dh} \quad . \tag{3.46}$$

Since $dh/ds \approx \varepsilon$ for small ε, the term d^2h/ds^2 describes the change of the local elevation with s. Hence, it is clear from Eq. (3.45) that if $d\mathsf{n}/dh = -1/R_E$, then dh/ds is constant and equal to zero if $\varepsilon = 0$. That is, the ray spreads parallel to the Earth's surface and the curvature of the ray path is $1/R_E$, but its curvature relative to the earth is zero. With Eq. (3.46) we can conclude that the ray's curvature C_E relative to the earth is

$$\mathsf{C}_E = -\frac{1}{R_E} - \frac{d\mathsf{n}}{dh} \quad . \tag{3.47}$$

3.5.1. Refractive index

The variation of n is closely related to the vertical variation of temperature T, water vapor partial pressure e and total pressure p. As n is slightly larger than unity (e.g., 1.0003), it is much more convenient to define the so-called refractivity N, given by

$$\mathsf{N} = (\mathsf{n} - 1) \times 10^6 \quad . \tag{3.48}$$

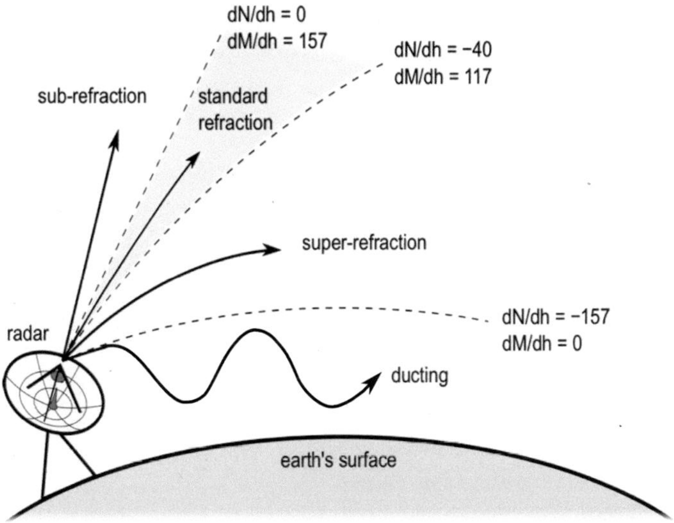

Fig. 3.8.: Common classification of atmospheric refraction conditions (in analogy to Turton et al. (1988))

For instance, if n is 1.0003, the corresponding value of N is 300. Bean and Dutton (1966) showed that N can be empirically given by

$$\mathsf{N} = c_1 \frac{p}{T} + c_2 \frac{e}{T} + c_3 \frac{e}{T^2} \quad , \tag{3.49}$$

where $c_1 = 77.6\,\mathrm{K\,hPa^{-1}}$, $c_2 = -6.0\,\mathrm{K\,hPa^{-1}}$, $c_3 = 3.75 \times 10^5\,\mathrm{K^2\,hPa^{-1}}$.

Normally, the refractivity N decreases with height in the atmosphere, which leads to a downward bending of radar beams. Under some circumstances, N may increase with the height, i.e., $d\mathsf{N}/dh > 0$, and the beam bends away from the Earth's surface. As mentioned above, a horizontal ray ($dh/ds = 0$) remains horizontal (i.e., has the same curvature as the earth), if $d\mathsf{n}/dh = -1/R_E$ resp. $d\mathsf{N}/dh = -10^6/R_E$, and a non-horizontal ray would preserve its local elevation, e.g., if $dh/ds > 0$, a quasi-helical motion around the earth would result. With $R_E \approx 6371$ km, this is $d\mathsf{N}/dh = -157\,\mathrm{km^{-1}}$. If the derivative of N is smaller than that, the curvature of the ray becomes larger than that of the earth, and the ray will, after reaching a maximum height, be bent down and trapped between this height and the Earth's surface. This process is called trapping, and the layer of the atmosphere within which the beam is bent back downwards is called trapping layer. If there is a region below the trapping layer with a larger derivative of N, the mode of beam propagation is similar to that of a waveguide, and this configuration is called a duct.

For the lowest 1 km of the ICAO (International Civil Aviation Organization) standard atmosphere, Doviak and Zrnic (1993) give a value of $d\mathsf{N}/dh = -40\,\mathrm{km}^{-1}$, and we will refer to this as "normal" conditions in the following, as they represent some "average" climatological conditions near the ground.

If $d\mathsf{N}/dh$ lies between 0 and -40 km^{-1}, the beam will be bent towards the Earth's surface with a curvature less than that of the normal conditions, and we refer to it as sub-refraction (see Fig. 3.8). Super-refraction occurs when $d\mathsf{N}/dh$ ranges from -40 to -157 km^{-1}. In this situation the beam is bent down to the surface at a rate less than the Earth's curvature but more than normal.

When considering atmospheric ducts, instead of N the so-called modified refractivity M is usually preferred, defined as

$$\mathsf{M} = \mathsf{N} + \frac{h}{R_E \times 10^{-6}} \approx \mathsf{N} + h\,157\,\mathrm{km}^{-1} \quad . \tag{3.50}$$

Then $d\mathsf{M}/dh = d\mathsf{N}/dh + 157\,\mathrm{km}^{-1}$ and for constant M ($d\mathsf{M}/dh = 0$) the curvature of the propagation of a nearly horizontal beam is that of the Earth's surface, and $d\mathsf{M}/dh < 0$ for trapping conditions. Fig. 3.8 shows the various categories of refraction in terms of $d\mathsf{N}/dh$ and $d\mathsf{M}/dh$.

According to the profile of M, three basic forms of a duct with corresponding duct depths are shown in Fig. 3.9. The case in Fig. 3.9(a) illustrates the structure associated with a simple surface duct. Here the duct extends from the local minimum to the surface, and the trapping layer, where $d\mathsf{M}/dh < 0$, stretches throughout the duct. Fig. 3.9(b) is referred to as the surface S-shaped duct, which reaches down to the surface, while the trapping layer doesn't, since $d\mathsf{M}/dh > 0$ near the surface. In these two cases, the duct depth is the height difference between the ground and the top of the duct where the minimum in modified refractivity profile is achieved. In Fig. 3.9(c), the common conditions for an elevated duct are given, where the value of M at the surface is less than that at the top of the duct, and so the duct can not reach down to

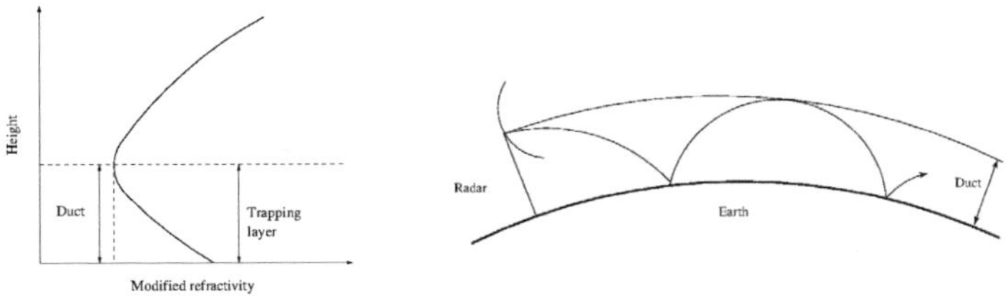

(a) Simple surface duct

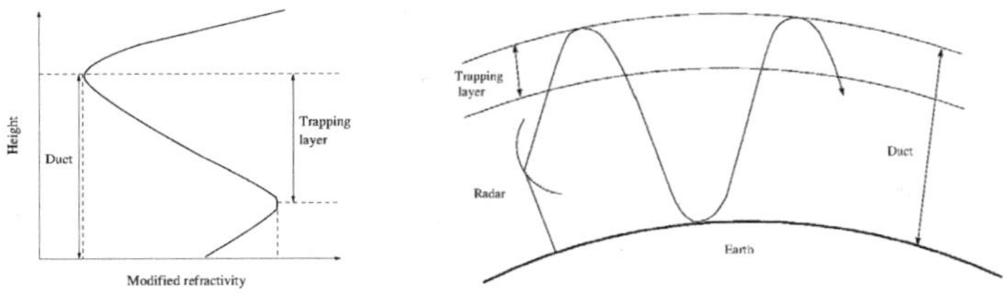

(b) Surface S-shaped duct

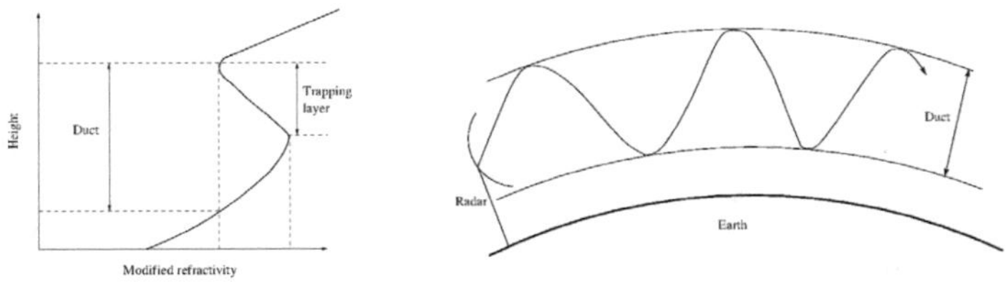

(c) Elevated duct

Fig. 3.9.: Left column: Typical modified refractivity M profiles. The depth of the ducts and the trapping layers are illustrated. **Right column:** Corresponding typical radar beam propagation paths in these ducting conditions (in analogy to Turton et al. (1988)).

the surface. Its depth extends from the local minimum to the height at which the M value equals that at the top of the duct.

As mentioned above, a duct is the result of strong vertical changes in the refractive index of the atmosphere between air masses of different temperatures and humidities, especially at low levels of the atmosphere. A duct can occur on a very large scale when a large mass of cold air is overrun by warm air, leading to a strong temperature inversion. A duct can also occur when a strong cap of warm and dry air exists in the lower troposphere above very moist air. On one hand, a duct causes the electromagnetic energy to be able to propagate over further distances, allowing long-range radio communication; on the other hand, in weather radar applications, ducts usually lead to coverage fades, increased ground clutter, increased anomalous propagation and range-height errors. One part of the radar simulation process is the computation of beam propagation within the atmosphere simulated by an NWP model in an appropriate way. It is known that low elevations are often vulnerable to anomalous propagation and orographic beam blockage, which can seriously affect the radar's ability to detect and quantify precipitation at ground level. Important issues here are to minimize the influences of these effects in the observation-simulation-comparison. In the following sections, we briefly describe and analyze a simple well-known approximate technique and two more sophisticated (new) methods.

3.5.2. 4/3 Earth radius Model

For convenience of computation, one likes to consider the ray path as a straight line. This can be accomplished by multiplying Eq. (3.47) with -1, resulting in

$$R_{eff} = -\frac{1}{\mathsf{C}_E} = \frac{1}{\frac{1}{R_E} + \frac{d\mathsf{n}}{dh}} = \underbrace{\frac{1}{1 + R_E \frac{d\mathsf{n}}{dh}}}_{=:K_{eff}} \cdot R_E \;=\; K_{eff} \cdot R_E \quad . \tag{3.51}$$

Here, R_{eff} denotes the Earth's curvature relative to a straight ray and K_{eff} is the effective Earth radius factor depending on $d\mathsf{n}/dh$. From Eq. (3.51) we can see that if $d\mathsf{n}/dh$ is constant, then the earth has an effective radius of constant R_{eff}.

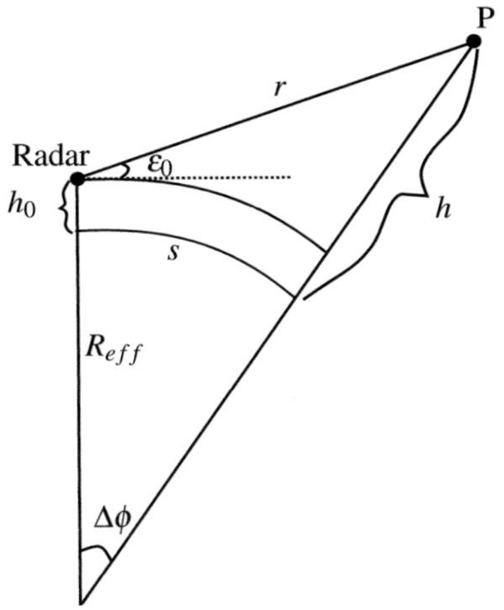

Fig. 3.10.: P is the position of ray at range r; $R_{eff} = 4/3R_E$; h_0 is the height of radar above MSL and ε_0 is the initial elevation of ray at radar antenna. Due to the geometric relationship, it holds $(h+R_{eff})^2 = (R_{eff}+h_0)^2 + r^2 - 2(R_{eff}+h_0)r\cos(90^\circ+\varepsilon_0)$ and $R_{eff}\Delta\phi = s$, where $\sin\Delta\phi = \dfrac{r\sin(90^\circ+\varepsilon_0)}{R_{eff}+h}$ due to the Law of Sines.

In the standard atmosphere, where the refractive index decreases linearly with the height in the mean by $d\mathsf{n}/dh = -40\times10^{-6}\,\text{km}^{-1}$ in the lowest 1 km or so (i.e., -40 N-units km^{-1} or 117 M-units km^{-1}), then it yields

$$K_{eff} = \frac{1}{1+R_E\dfrac{d\mathsf{n}}{dh}} \approx \frac{4}{3} \quad . \tag{3.52}$$

This is a common model to approximate ray paths, which assumes that the effective Earth radius is 1/3 larger than the real one, so $R_{eff} = 4/3R_E$. This model allows for a straightforward analytical estimation of each pulse volume height h and surface distance s relative to the radar site at a height of $h = 0$ (MSL) by (cf. Fig. 3.10):

$$h = h(R_{eff},\varepsilon_0) = \sqrt{(R_{eff}+h_0)^2 + r^2 + 2(R_{eff}+h_0)r\sin\varepsilon_0} - R_{eff} \quad , \tag{3.53}$$

$$s = s(R_{eff},\varepsilon_0) = R_{eff}\arcsin\left(\frac{r\cos\varepsilon_0}{R_{eff}+h}\right) \quad . \tag{3.54}$$

This model is referred to as the "4/3 Earth radius model" and abbreviated in the following by 43ERM. As shown in Doviak and Zrnic (1993), for weather radar applications 43ERM can be used for all elevations, if h is confined to the first 10-20 km and if n has a slope of $-1/(4R_E)$ in the first kilometer of the atmosphere. But the slope of n is usually not constant. If n decreases much more rapidly than in the standard atmosphere, the beam will likely be bent downwards, and then the height of pulse volumes tends to be overestimated by 43ERM. These errors can be quite significant for elevations smaller than $\sim 1°$. More sophisticated methods can be found in Caumont (2006), Gao et al. (2006), Chen et al. (2009) and Siebren (2003), however, their performances and limitations, especially under the challenging ducting conditions for low elevations, have been rigorously reviewed to date. Therefore, two robust methods are introduced in the following.

3.5.3. Method based on the total reflection

Under realistic atmospheric conditions, e.g., n-profile based on a radiosonde measurement, some authors computed the ray propagation iteratively by discretizing Snell's law Eq. (3.43) in the along-beam direction r in steps of some fixed increment Δr.

In this sense, a method used in Caumont (2006) has been revised, adding a simple criterion for total reflection.

Let $l = 1, \ldots, L$ be the numeration index of steps. Then the height h_l and the MSL reduced surface distance s_l at some location are iteratively calculated from the values at $l-1$ (height above MSL h_{l-1}, local elevation ε_{l-1}) under the assumption of straight rays within Δr

$$h_l = \sqrt{(R_E + h_{l-1})^2 + \Delta r^2 + 2(R_E + h_{l-1})\Delta r \sin\varepsilon_{l-1}} - R_E \quad , \tag{3.55}$$

$$s_l = s_{l-1} + \Delta s_l = s_{l-1} + R_E \Delta\phi_l = s_{l-1} + R_E \arcsin\left(\frac{\cos\varepsilon_{l-1}\,\Delta r}{R_E + h_l}\right) \quad . \tag{3.56}$$

Fig. 3.11 shows a sketch of these quantities.

Given the new h_l, the new ε_l can be directly derived from the discretized Eq. (3.43),

$$\varepsilon_l = \pm\arccos \underbrace{\left(\frac{(R_E + h_{l-1})}{(R_E + h_l)} \frac{\mathsf{n}_{l-1}}{\mathsf{n}_l} \cos\varepsilon_{l-1}\right)}_{:=F} \quad . \tag{3.57}$$

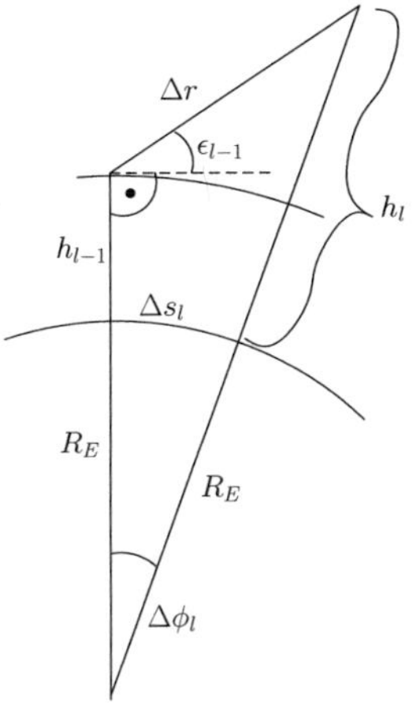

Fig. 3.11.: Sketch of a straight ray path segment Δr and corresponding MSL reduced arc segment Δs_{l-1} for the TORE method.

where ε_l is the local elevation of the ray at range $l\Delta r$.

Two problems arise here:

(1) arccos is not defined in the case of $F > 1$;

(2) the sign of ε_l is ambiguous ($\pm$) because arccos is not a unique mapping for the co-domain $[-\frac{\pi}{2}, \frac{\pi}{2}]$.

Concerning (1), this could physically happen if n_l is "sufficiently" smaller than n_{l-1} at some location l . In analogy to total reflection at a discrete n-jump, we here assume that the ray be reflected back internally, so

$$\varepsilon_l = -\left(\varepsilon_{l-1} + \underbrace{\frac{\Delta r \cos \varepsilon_{l-1}}{R_E + h_{l-1}}}_{=:\Delta\varepsilon}\right) \quad . \tag{3.58}$$

$\Delta\varepsilon$ is a correction term which accounts for the effect of the Earth's curvature on the local elevation along a Δr-segment. A graphical derivation of $\Delta\varepsilon$ can be found in Fig. 3.12.

If $F \leq 1$, it is reasonable to assume for starters that the sign of ε does not change from one step to the next, so we have instead of Eq. (3.57)

$$\varepsilon_l = \begin{cases} \mathrm{sign}(\varepsilon_{l-1}) \arccos(F) & \text{,if } F \leq 1; \\ -(\varepsilon_{l-1} + \Delta\varepsilon) & \text{,otherwise} \end{cases} \quad . \tag{3.59}$$

At first glance, the criterion $F \leq 1$ or $F > 1$ in the iteration of Eqs. (3.55) and (3.59) could also work for the case of negative elevations, where the sign changes from - to + somewhere along the ray under "normal" propagation conditions due to the Earth's curvature. In this case, it holds $h_l < h_{l-1}$ in F, so that $F > 1$ would be theoretically possible. But we found by extensive testing that the asymptotic behavior of F as a whole is such that it usually remains ≤ 1 in the iteration when ε_{l-1} approaches 0 from the negative side (except for cases where the n-gradient is very weak), and no sign change occurs. An example can be found in the later Experiment 4 in Section 3.5.5. Unfortunately, this asymptotic behaviour seems to be independent of the choice of Δr. To circumvent this problem, an extra ad hoc criterion is adopted, which uses the increment between ε_{l-1} and ε_{l-3} to linearly extrapolate and predict ε_l. A sign change is assumed if $\varepsilon_{l-1} < 0$ and $\varepsilon_{l-1} + (\varepsilon_{l-1} - \varepsilon_{l-3}) > 0$. The reason we choose the increment between ε_{l-1} and ε_{l-3} instead of the increment between ε_{l-1} and ε_{l-2} is that the later one is often too small to prompt the sign change, again because of the asymptotic behavior of the iteration when ε_{l-1} approaches 0 from the negative side. This ad hoc criterion works effectively as shown in the Experiment 4 in Section 3.5.5.

Therefore, Eq. (3.59) is modified to

$$\varepsilon_l = \begin{cases} -\,\mathrm{sign}\,(\varepsilon_{l-1})\arccos F & \text{,if } F \leq 1 \wedge (\varepsilon_{l-1} < 0 \wedge \varepsilon_{l-1} + (\varepsilon_{l-1} - \varepsilon_{l-3})) > 0\,; \\ -(\varepsilon_{l-1} + \Delta\varepsilon) & \text{,if } F > 1\,; \\ \mathrm{sign}(\varepsilon_{l-1}) \arccos F & \text{,otherwise} \end{cases} \quad . \tag{3.60}$$

Because of the newly considered total reflection assumption, this modified method is called TORE (acronym for TOtal REflection) and can be summarized as follows (see Fig. 3.13):

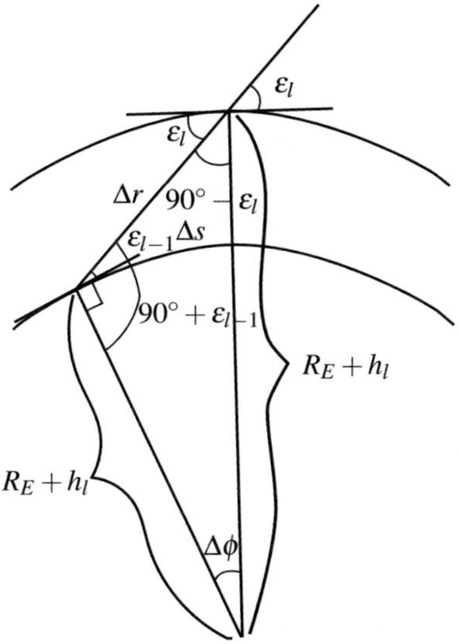

Fig. 3.12.: Derivation of $\Delta\varepsilon$: on one hand, because $(R_E + h_{l-1})\Delta\phi = \Delta s$, it holds $\Delta\phi = \frac{\Delta s}{R_E + h_{l-1}} = \frac{\Delta r \cos \varepsilon_{l-1}}{R_E + h_{l-1}}$. On the other hand, it holds $(90° - \varepsilon_l) + (90° + \varepsilon_{l-1}) + \Delta\phi = 180°$. Together, it yields $\Delta\varepsilon = \varepsilon_l - \varepsilon_{l-1} = \Delta\phi = \frac{\Delta r \cos \varepsilon_{l-1}}{R_E + h_{l-1}}$.

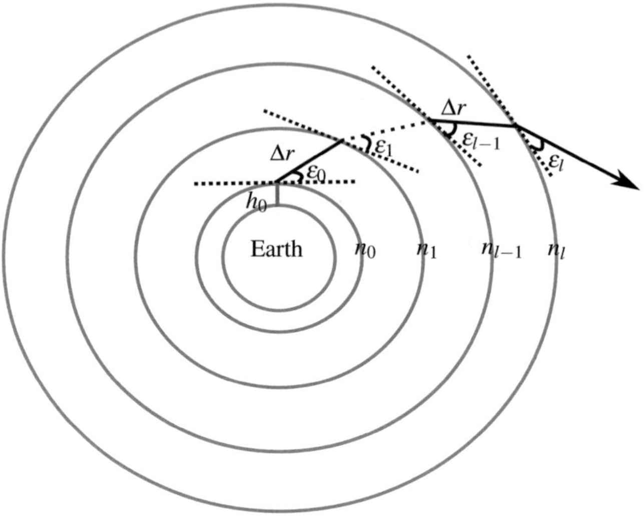

Fig. 3.13.: Illustration of the iterative polygon pieces of length Δr and successive local elevations ε_l and refractive indices n_l for the TORE method.

Step 1: Calculate height h_l and MSL-reduced surface distance s_l using Eqs. (3.55-3.57), starting from h_{l-1} and s_{l-1} with local elevation ε_{l-1},

Step 2: Estimate $n_l = n(h_l)$ using radiosonde or NWP data,

Step 3: Calculate ε_l using Eq. (3.60). Note that n_{l-1} is known from Step 2 of the previous iteration.

Steps 1 - 3 are repeated from $l = 1$ to $l = L$. In the first iteration, the values at $l - 1$ are antenna elevation ε_0, height h_0 at the radar antenna, refractive index $n_0 = n(h_o)$ and $s_0 = 0$.

Note that, despite extensive testing, we cannot exclude that the above ad hoc criterion might fail in rare instances, because it is not rigorously mathematically well-founded. Note also, that the above sign ambiguity is a general problem of Snell's law (as stated earlier), and that all methods based on it have to deal with the problem in one way or another.

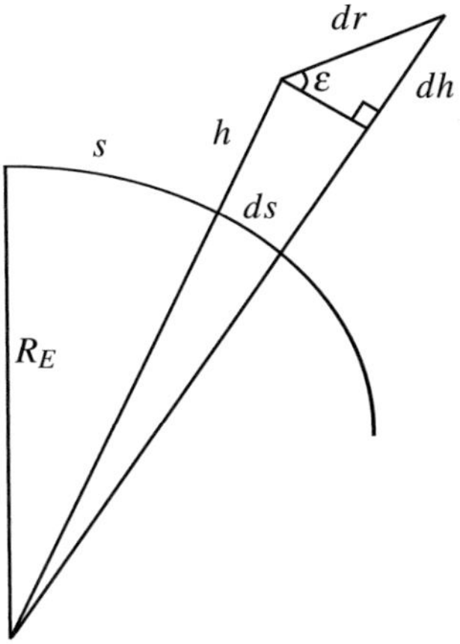

Fig. 3.14.: Sketch of R_E, h, s and their differentials for the derivation of Eqs. (3.68) and (3.69)

3.5.4. Method using Second-order Ordinary Differential Equation

Although TORE considers explicitly the actual refractive index, an ad hoc criterion is required to determine the sign change of local elevations. In this section, we have found a novel method, called SODE (abbreviation for Second-order Ordinary Differential Equation), which offers a straightforward analytical/numerical solution for ray propagation and considers the sign change automatically.

As can be seen from Eq. (3.42), the ray propagation can be formulated as an initial value problem of an ODE. Principally this would be possible by employing Eq. (3.42) directly, but it has the drawback of being formulated relative to the independent coordinate s. For many practical applications, e.g., as part of a radar forward operator, a formulation relative to the along-beam range r would be preferential, because it is then possible to discretize the solution by using a constant Δr which is directly related to the radar range.

To derive such an alternative ODE, we start from Snell's law Eq. (3.43) but now assume h as a function of r, i.e.,

$$\mathsf{n}(h(r))\,[R_E + h(r)]\cos\varepsilon \;=\; const \quad . \tag{3.61}$$

From differential geometry (cf. Fig. 3.14) one obtains for infinitely small dh and dr

$$\sin\varepsilon = \frac{dh}{dr} \tag{3.62}$$

$$\cos\varepsilon = \sqrt{1-\left(\frac{dh}{dr}\right)^2} \quad \left(\leq 0 \quad \text{because of } \varepsilon \in [-\frac{\pi}{2},\frac{\pi}{2}]\right) \quad , \tag{3.63}$$

therefore

$$\mathsf{n}(h(r))\,[R_E+h(r)]\sqrt{1-\left(\frac{dh}{dr}\right)^2} = const \quad . \tag{3.64}$$

As indicated, h is assumed to be a function of the range r, so the refractive index n depends implicitly on r. One differentiates Eq. (3.64) with respect to r and obtains

$$\frac{d\mathsf{n}}{dh}\frac{dh}{dr}(R_E+h)\sqrt{1-\left(\frac{dh}{dr}\right)^2}+\mathsf{n}\frac{dh}{dr}\sqrt{1-\left(\frac{dh}{dr}\right)^2}+\mathsf{n}\,(R_E+h)\frac{-2\frac{dh}{dr}\frac{d^2h}{dr^2}}{2\sqrt{1-\left(\frac{dh}{dr}\right)^2}}=0 \quad , \tag{3.65}$$

The singularities $dh/dr=\pm 1$ can be removed by multiplying $\sqrt{1-(dh/dr)^2}$, which yields

$$\frac{dh}{dr}\left\{\mathsf{n}(R_E+h)\frac{d^2h}{dr^2}+\left(\frac{dh}{dr}\right)^2\left[\frac{d\mathsf{n}}{dh}(R_E+h)+\mathsf{n}\right]-\left[\frac{d\mathsf{n}}{dh}(R_E+h)+\mathsf{n}\right]\right\}=0 \quad . \tag{3.66}$$

Since $dh/dr=0$ as a general solution is unphysical, it holds

$$\mathsf{n}(R_E+h)\frac{d^2h}{dr^2}+\left(\frac{dh}{dr}\right)^2\left[\frac{d\mathsf{n}}{dh}(R_E+h)+\mathsf{n}\right]-\left[\frac{d\mathsf{n}}{dh}(R_E+h)+\mathsf{n}\right]=0 \quad . \tag{3.67}$$

One divides both side of Eq. (3.67) by $\mathsf{n}(R_E+h)$ and obtains the second-order nonlinear ODE

$$\frac{d^2h}{dr^2}+\left(\frac{dh}{dr}\right)^2\left(\frac{1}{\mathsf{n}}\frac{d\mathsf{n}}{dh}+\frac{1}{R_E+h}\right)-\left(\frac{1}{\mathsf{n}}\frac{d\mathsf{n}}{dh}+\frac{1}{R_E+h}\right)=0 \tag{3.68}$$

and, by substituting $dh/dr = u$, the equivalent set of two coupled first-order equations are arrived at

$$\begin{aligned} \frac{dh}{dr} &= u \quad , \\ \frac{du}{dr} &= -u^2\left(\frac{1}{\mathsf{n}}\frac{d\mathsf{n}}{dh} + \frac{1}{R_E+h}\right) + \left(\frac{1}{\mathsf{n}}\frac{d\mathsf{n}}{dh} + \frac{1}{R_E+h}\right) \quad . \end{aligned} \tag{3.69}$$

Eqs. (3.68-3.69) are physically equivalent to Eq. (3.42), but are formulated with independent coordinate r instead of s. It can be easily checked that Eqs. (3.68-3.69) also include the constant solution $dh/dr = 0$ describing a ray along the Earth's surface for $d\mathsf{n}/dh = -\mathsf{n}/(R_E + h)$ and the cases $dh/dr = \pm 1$ for the exact vertical propagation (an advantage against Eq. (3.42)).

The equation system (3.69) can be treated as an initial value problem (IVP) with initial values

$$u(r=0) = \left.\frac{dh}{dr}\right|_{r=0} = \sin\varepsilon_0 \quad , \tag{3.70}$$

$$h(r=0) = h_0 \quad , \tag{3.71}$$

and the ray tracing problem is then uniquely solved by this IVP.

In analogy to TORE, Eq. (3.69) is discretized and solved in steps of Δr. The iteration step from location $l-1$ to l is done as follows:

Step 1: Estimate $1/\mathsf{n}_{l-1}$ and $d\mathsf{n}/dh|_{l-1}$ at the height h_{l-1} using radiosonde soundings or NWP data,

Step 2: Solve Eq. (3.69) with initial values u_{l-1} and h_{l-1} to obtain u_l and h_l

Step 3: As in Eq. (3.57), calculate MSL-reduced surface distance s_l from

$$s_l = s_{l-1} + R_E \arcsin\left(\frac{\cos\varepsilon_{l-1}\,\Delta r}{R_E + h_l}\right) \quad ,$$

with

$$\varepsilon_{l-1} = \arcsin\left(u_{l-1}\right) \quad . \tag{3.72}$$

Steps 1 - 3 are repeated from $l = 1$ to $l = L$. Note that the IVP posed in Step 2 is currently solved using the 4th order explicit Runge-Kutta method (RK4, see Appendix B.2), but any other numerical standard method for solving ODEs would be suitable as well. For the first iteration $l = 1$, the initial values in Step 2 are given by Eqs. (3.70-3.71), and the refractive index and its slope has to be estimated at h_0, resulting in $1/\mathsf{n}_0$ and $d\mathsf{n}/dh|_0$.

Other authors have also applied differential equation solvers for the ray propagation problem (Hartree et al., 1946; Siebren, 2003), but the formulation of the above ODE in terms of r is believed to be new and especially suitable for radar forward operators.

3.5.5. Sensitivity experiments

So far we have presented three methods to calculate radar beam propagation in a stratified atmosphere. In what follows, we compare all these methods by evaluating them for specific atmospheric conditions. This is done with a series of sensitivity experiments in a framework, where certain horizontally homogeneous vertical profiles of T, p and e are prescribed (cf. Neuper (2010)). The first three experiments are based on the idealized ducting profiles introduced in Fig. 3.9. A fourth experiment is based on standard atmosphere data, and a fifth applies measured radiosonde data from a ducting case.

For all experiments we choose a maximal surface cover range of 300 km and a range resolution Δr of 500 m. In the first three experiments we investigate simulations of beam propagation for two initial elevations $\varepsilon_0 = 0.1°$ and $\varepsilon_0 = 1.1°$, and in the fourth one for $\varepsilon_0 = -0.3°$. In order to stimulate different kinds of ducts, the radar antenna is set accordingly to different heights in the experiments.

Experiment 1: idealized surface duct

In this experimental setup we simulate a surface duct. Accordingly, we have chosen the profiles of M and N with respect to h as given in Figs. 3.15(a) and 3.15(b), respectively. The radar antenna is set at a height of 200 m. Fig. 3.15(a) shows a large negative slope of -100 M-units km^{-1} of M for the first 350 m and thereafter a slope of 117 M-units km^{-1}.

The simulation results are shown in Figs. 3.15(c) and 3.15(d). Fig. 3.15(c) represents the variations of beam heights h with distance s computed by the three methods, while Fig. 3.15(d) shows the absolute height differences of TORE and SODE, respectively, compared to 43ERM. One can see that for $\varepsilon_0 = 1.1°$, the beam heights calculated by all three methods are generally close to each other (as shown in Fig. 3.15(d)) and are less than 600 m at maximum range; for $\varepsilon_0 = 0.1°$, the resulting rays according to TORE and SODE are refracted downward to the Earth's surface because of the strong negative $d\mathsf{M}/dh$, while 43ERM produces a curve which is straightening up. No surface reflection was taken into account in these calculations, and therefore the rays of TORE and SODE end when reaching the ground. The discrepancies compared to 43ERM grow already to about 1600 m at a distance of 140 km. This observation is therefore in accordance with the statement from Doviak and Zrnic (1993) that for the higher elevations the radar beams are less sensitive to the refractivity gradient, while for low elevations ($< 1°$) under ducting conditions 43ERM is prone to (strongly) overestimating the beam heights.

Experiment 2: idealized surface S-shaped duct

Now we apply all these methods to an idealized surface S-shaped duct, characterized by the profiles of M and N, shown in Figs. 3.16(a) and 3.16(b). The antenna's height is chosen to 40 m. The profile of M begins with a slope of 117 M-units km^{-1} for the lowest 100 m, then alters to -100 M-units km^{-1} until 400 m and thereafter goes back to 117 M-units km^{-1}. As can be seen in Figs. 3.16(c) and 3.16(d), for $\varepsilon_0 = 1.1°$ the differences of three methods are insignificant (about 1 km at maximal distance), in contrast to $\varepsilon_0 = 0.1°$, where the rays calculated by SODE and TORE are both propagated in a wave-like mode, trapped within the ducting layer (see Fig. 3.16(c)), and the height differences compared to 43ERM reach 1600 m at a distance of 150 km (see Fig. 3.16(d)). The reason for slight discrepancies between SODE and TORE for $\varepsilon_0 = 0.1°$ in Fig. 3.16(c) will be addressed in Experiment 5.

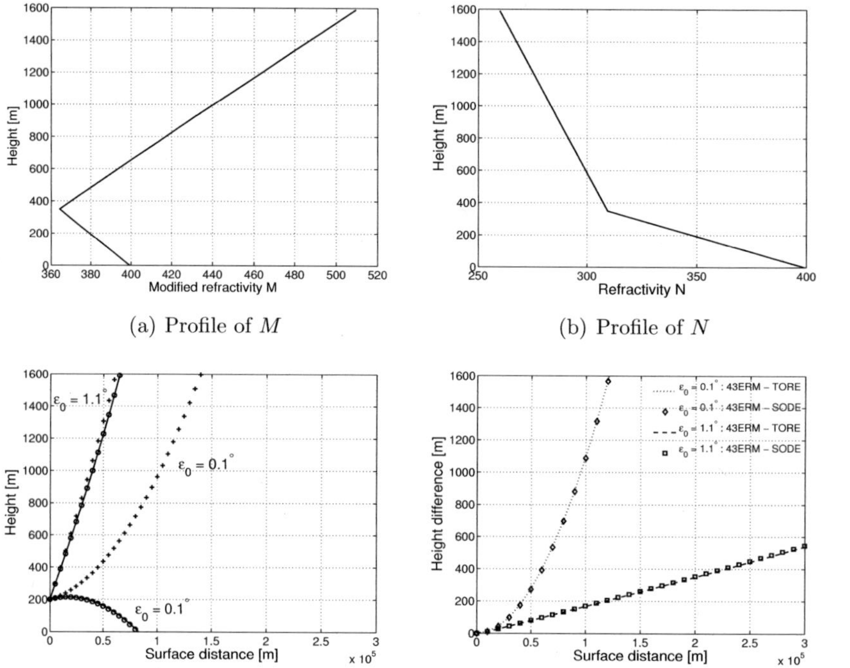

(a) Profile of M

(b) Profile of N

(c) Simulation using three methods: + 43ERM; — SODE; o TORE

(d) Height differences in reference to 43ERM: 43ERM - TORE: beam heights computed by 43ERM subtracted by those of TORE; analogously for 43ERM - SODE

Fig. 3.15.: Experiment 1: simple surface duct. (a) and (b): profiles of M and N with height in m; (c): beam heights in m for different initial elevations and simulation methods as indicated as a function of surface distance in m$\times10^5$. The antenna height is assumed to be 200 m; (d): height differences compared to 43ERM as a function of surface distance in m$\times10^5$.

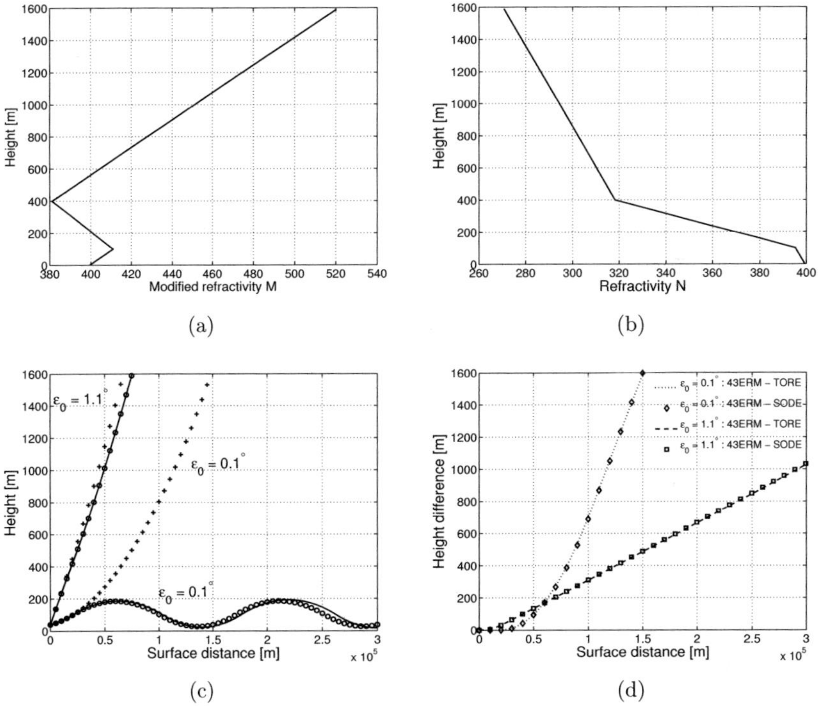

Fig. 3.16.: Experiment 2: same as Fig. 3.15 but for a surface S-shaped duct

Experiment 3: idealized elevated duct

A further situation modelled is an idealized elevated duct. The corresponding profiles of M and N are illustrated in Figs. 3.17(a) and 3.17(b). The antenna height is 300 m. The profile of M starts with slope 117 M-units km^{-1} for the first 250 m, then changes to -100 M-units km^{-1} until 400 m and at last returns to 117 M-units km^{-1}. The general features of the results, illustrated in Figs. 3.17(c) and 3.17(d), are mainly the same as those in Experiment 2 except that the duct here is lifted in the air and does not touch the ground.

Experiment 4: standard conditions

The current experiment is based on data for standard conditions. As shown in Fig. 3.18(a) (or 3.18(b)), now a constant slope of $d\mathsf{M}/dh = 117$ M-units km^{-1} throughout the atmosphere is considered. The antenna height is set to 200 m. But now the elevation is set to a negative value, $\varepsilon_0 = -0.3°$. In order to demonstrate the effects of the ad hoc approach Eq. 3.59) for TORE, two experiments are performed here, one using Eq. (3.59), denoted with E4(1) and the other one simplily $\varepsilon_l = \mathrm{sign}(\varepsilon_{l-1}) \cdot \arccos(F)$, denoted with E4(2). In E4(1), the results of all three methods are basically identical (see Figs. 3.18(c) and 3.18(e)). All prompt the beams to descend at the beginning in virtue of the negative initial elevation and to slope upwards after a distance of about 50 km due to the Earth's curvature. This shows that, for "normal" conditions, 43ERM is a satisfactory approximation in comparison with the solution of SODE, which is considered as an accurate reference solution.

But in E4(2), TORE is not able to overcome the slight negative elevations (near to 0° as shown in Fig. 3.18(e)) and thus flattens out afterwards, which is due to the fact that, under standard conditions, the beam parts with negative local elevation are propagated from smaller to larger n. Thus, the conditions for "total reflection" are not met and the negative local elevations fail to become positive, because Eq. (3.60) preserves the sign of the elevation from one TORE iteration step to the next. This special artifact shows the necessity of Eq. 3.59) as a criterion for a sign change of ε in "non-total-reflection" conditions (i.e., the elevation at some distance has to change its sign from negative to positive just because of the Earth's curvature).

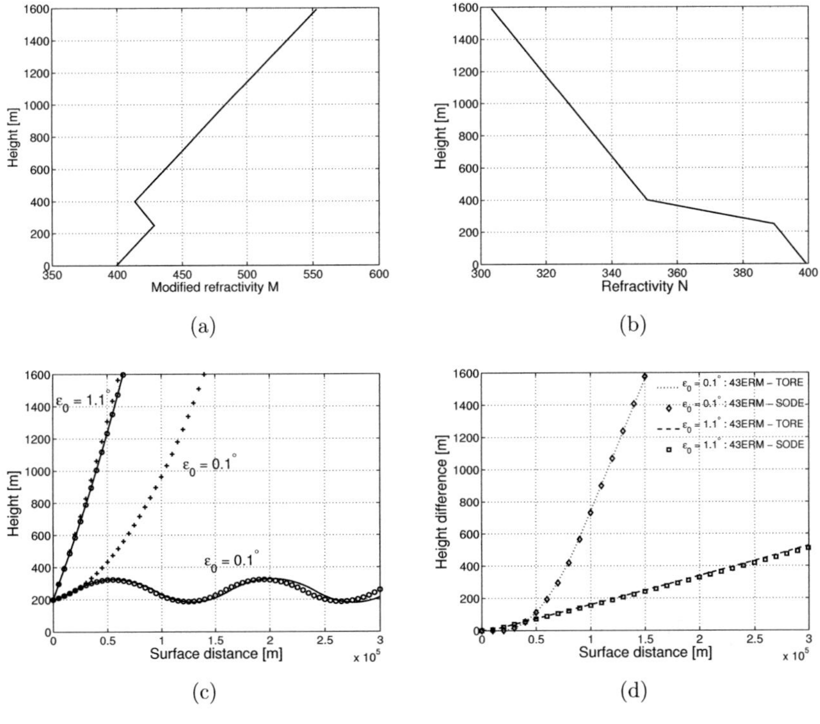

Fig. 3.17.: Experiment 3: same as Fig. 3.15 but for an elevated duct

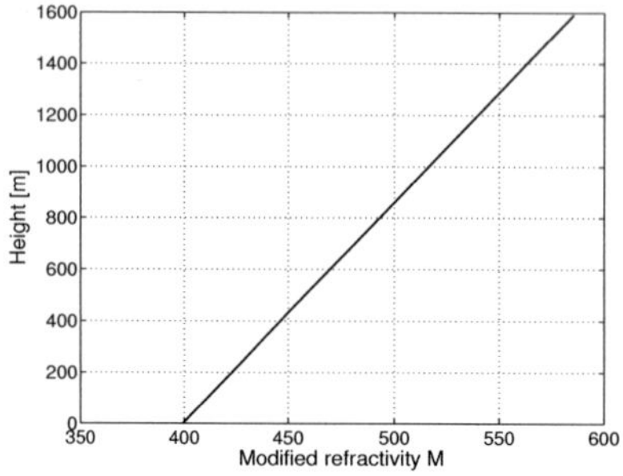

(a)

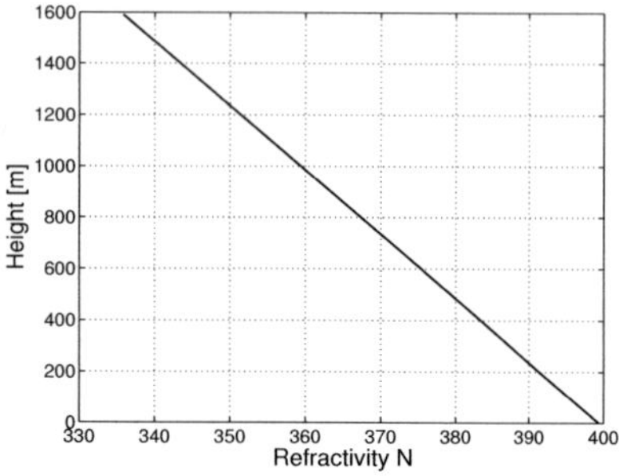

(b)

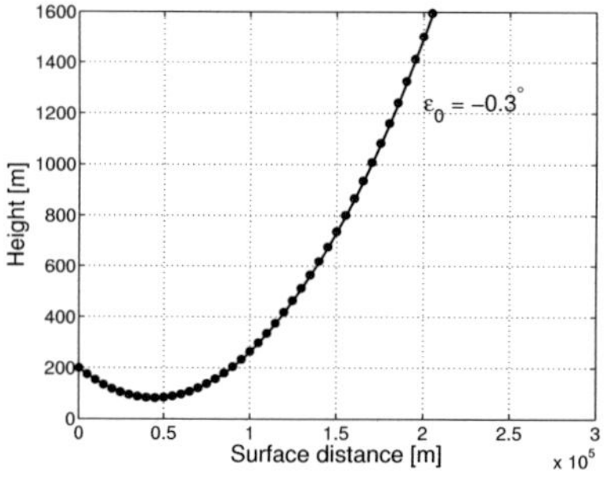

(c) E4(1)

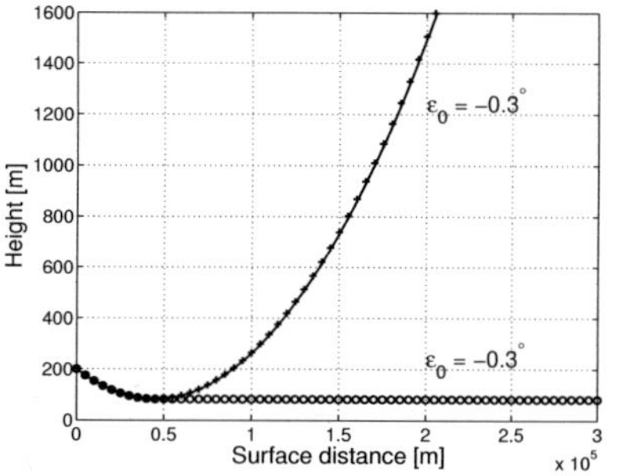

(d) E4(2)

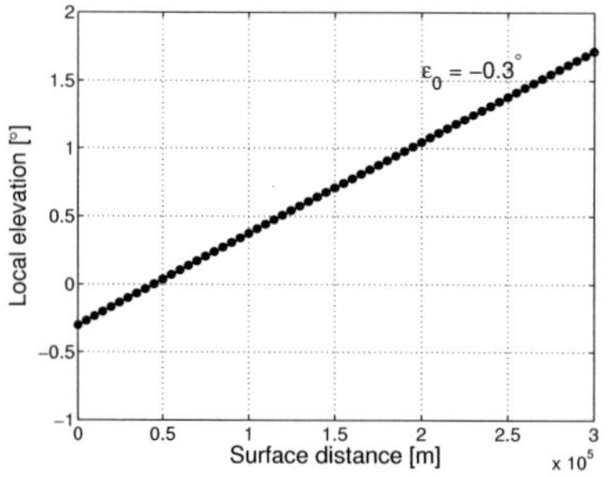

(e) E4(1): variations of local elevation as function of surface distance for elevation $-0.3°$: + 43ERM; — SODE; ∘ TORE

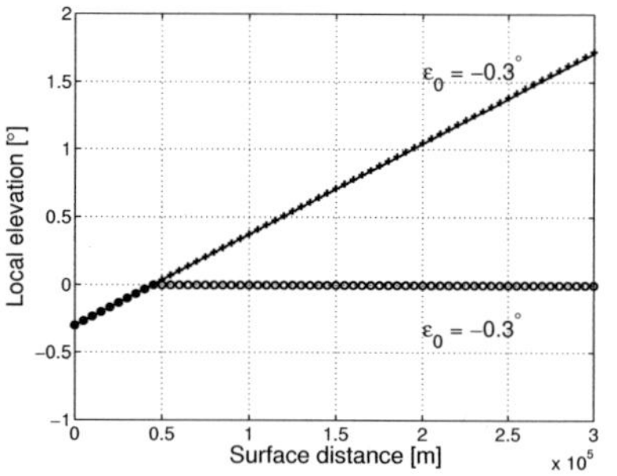

(f) As (e) but for E4(2)

Fig. 3.18.: Experiment 4: standard conditions. (c) and (d): beam heights in m calculated by TORE, using Eq. (3.59) or simplily $\varepsilon_l = \mathrm{sign}(\varepsilon_{l-1}) \cdot \arccos(F)$, respectively, indicated as a function of surface distance in m$\times 10^5$; (e) and (f): same as (c) and (d) but for the local elevation in degree.

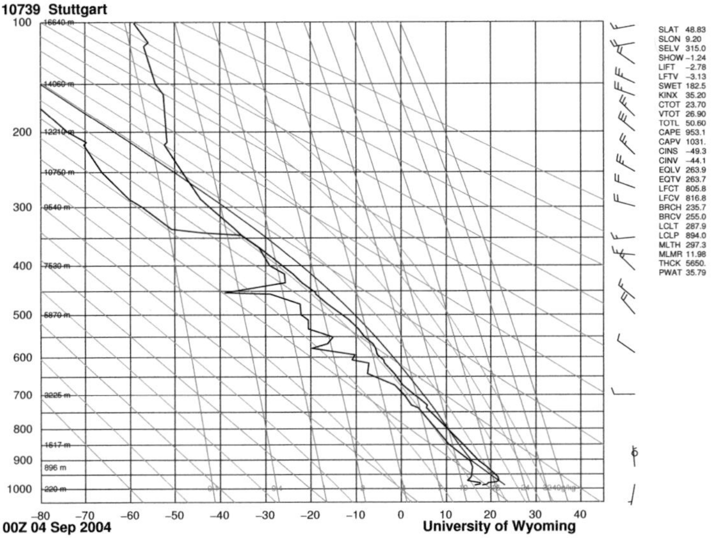

Fig. 3.19.: Sounding from Stuttgart-Schnarrenberg at 00:00 UTC on 4th September 2004

Experiment 5 using measured radiosonde data

In this experiment, a case based on real atmospheric conditions is investigated, which exhibits a strong temperature inversion and moisture profile near the ground observed at Stuttgart-Schnarrenberg (WMO-ID 10739) in Germany at 00 UTC on 4th September 2004 (see Fig. 3.19). The corresponding profiles of M and N given in Figs. 3.20(a) and 3.20(b) are derived from the radiosonde data available from the University of Wyoming[3], where a vertical interpolation of original T, dew point T_d and p-data to additional levels (linear oversampling every 10 m vertical) is performed and from these oversampled data, n is computed. The refractive index n_l at some arbitrary level l is derived by linear interpolation from upper and lower neighboring

[3] Although a radiosonde measures on a spatial scale much larger than the radar wavelength, the data are readily available and are a good source of atmospheric information on temperature and humidity structure. It is assumed that the radiosonde data will at least yield a not too noisy representation of the n-profile in ducting conditions.

oversampling points at locations $l_>$ and $l_<$. The refractivity slope is approximated by a simple differential quotient

$$\left.\frac{d\mathsf{n}}{dh}\right|_l = \frac{\mathsf{n}_{l_>} - \mathsf{n}_{l_<}}{h_{l_>} - h_{l_<}} \quad . \tag{3.73}$$

The linear oversampling of the original radiosonde data minimizes interpolation artifacts for n and its vertical slope. It is justified because in radiosonde data, T and T_d often exhibit a near-linear dependence on height inbetween the data points which are stored in radiosonde data sets, and p varies smoothly with height.

If, however, some noise in the derived n-profiles should lead to noisy gradients, some smoothing could be obtained by applying more sophisticated methods for interpolation and slope calculation. The authors found the so-called Savitzky-Golay-Filter (Press et al. (1993)) very useful, i.e., fitting of a low-order polynomial to a wider stencil of neighboring oversampling points (e.g., 5th order polynomial, 12 surrounding points) and computing n and its slope from this polynomial instead of interpolating from the original data.

The setup of maximal surface distance, range resolution and initial elevations remain the same as previously and the radar antenna height is set to 40 m.

The corresponding M-profile in Fig. 3.20(a) shows a duct between 17 m and 144 m above the surface (i.e., duct depth = 127 m) and a trapping layer extending from 71 m to 144 m. Therefore, it can be expected to observe an elevated duct between 17 m and 144 m in ray paths with low elevations. Fig. 3.20(c) illustrates the comparison of beam heights computed by the three methods. As expected, all three methods provide nearly the same results for $\varepsilon_0 = 1.1°$; for $\varepsilon_0 = 0.1°$, 43ERM, not "knowing" about the actual ducting conditions, generates a lifting curve, while TORE and SODE are consistent with the expected ducting conditions, and both are able to deliver reasonable waveguide-like results, with however slightly different shapes (Figs. 3.20(c) and 3.20(d)). Fig. 3.20(e) shows the corresponding local elevations.

It is worth noting that the discrepancies between SODE and TORE, as shown in Fig. 3.20(c), arise from numerical reasons, which can be eliminated by refining the range resolution Δr. For instance, if we replace $\Delta r = 500$ m with $\Delta r = 200$m, the discrepancies almost disappear as illustrated in Figs. 3.21(c) and 3.21(e).

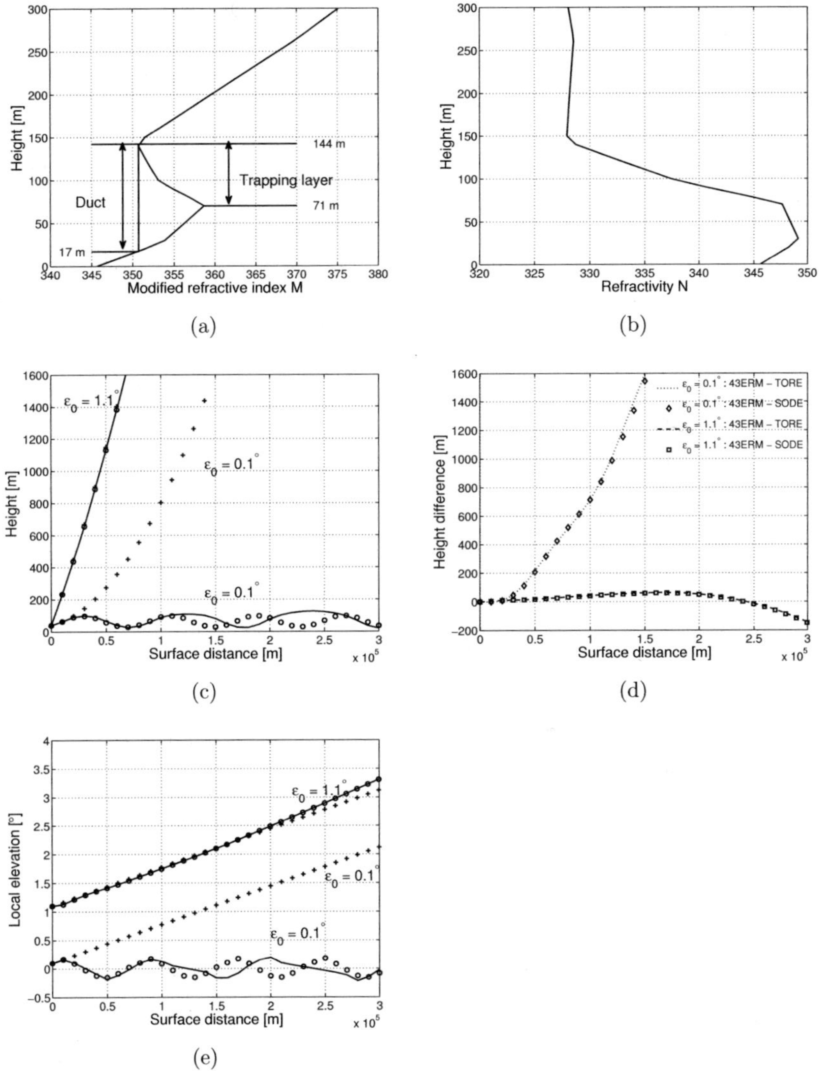

Fig. 3.20.: Experiment with real radiosonde data. (c): beam heights in m for different initial elevations and simulation methods as indicated as a function of surface distance in m$\times 10^5$. The antenna height is assumed to be 40 m; (d): height differences compared to 43ERM as a function of surface distance in m$\times 10^5$; (e): same as (c) but for the local elevation in degree.

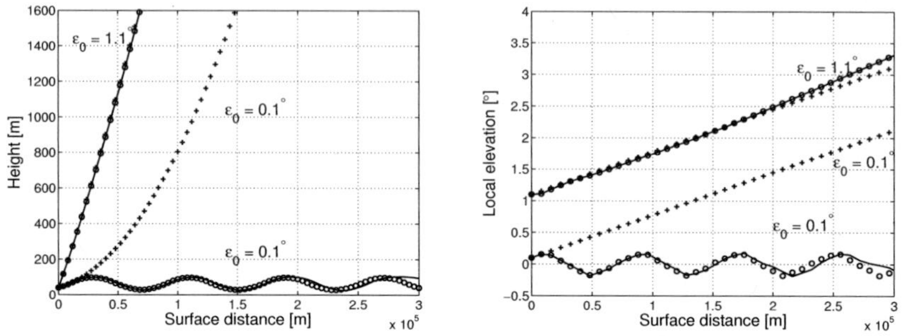

(a) (b)

Fig. 3.21.: As Fig 3.20 but with a finer range resolution of $\Delta r = 200$ m

Idealized case study with the COSMO-model

SODE, as the reference method, gives us an incentive to take a further insight into viability of SODE under abnormal conditions within the COSMO-model. Therefore, we test SODE again with the same thermodynamic profile given in Fig. 3.19 but now in framework of the COSMO-model. Here, special care must be given to the vertical grid resolution. Thus, two idealized COSMO-model runs are done with $exp\_galchen = 2.6$ and $exp\_galchen = 3.6$. Both grids have 64 vertical levels that are unequally spaced as shown in Tab. 3.2. Those values are interpolated, in accordance with the value of *exp_galchen*, from the initial thermodynamic profile to the model levels, and serve as initial data for model runs. We can see that the larger *exp_galchen* is, the denser the lower vertical levels are. Lateral boundaries are periodic. A radar station with an effective range of 124 km and an altitude of 100 m is assumed, thereby two elevations $\varepsilon_0 = 0.1°$ and $0.5°$ are investigated.

Vertical index	*exp_galchen* = 2.6		*exp_galchen* = 3.6	
	Height [m]	Temperature [K]	Height [m]	Temperature [K]
64	37.693	292.273	4.247	291.523
63	130.821	294.568	19.108	292.084
62	251.433	294.692	45.840	292.450
61	389.199	293.751	83.223	293.454
60	540.569	292.572	130.689	294.782
59	703.551	291.089	187.889	294.880
⋮	⋮	⋮	⋮	⋮
1	21717.759	218.345	21611.143	218.183

Tab. 3.2.: Interpolated vertical profiles of temperature on the first grid points of each model level: the vertical index begins from the top of the model. The heights result from the formula (Gal-chen and Somerville, 1975): $z = z^* \left(\frac{2}{\pi} \arccos\left(\frac{k-1}{ke-1}\right)\right)^{exp\_galchen}$, where z^* is the height of the model top level and *ke* is the total number of model main levels.

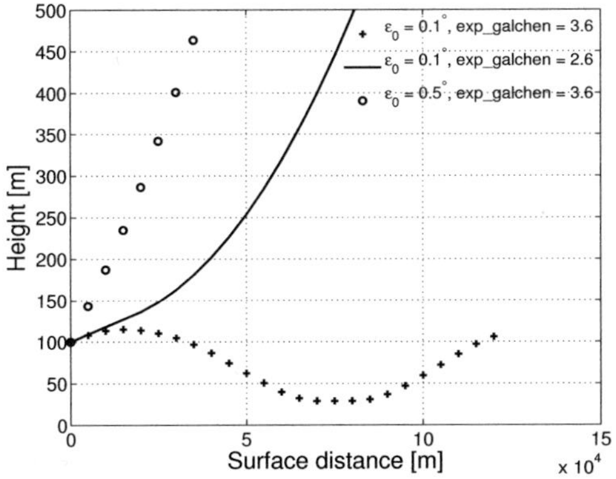

Fig. 3.22.: The idealized case study with the COSMO-model: beam heights in m, calculated by SODE for different initial elevations and exp_galchen, as indicated as a function of surface distance in m$\times 10^4$. The antenna height is assumed to be 100 m

The results of experiments are demonstrated in Fig. 3.22. With $exp\_galchen = 3.6$, a wave-shaped ray path for $\varepsilon_0 = 0.1°$ is produced, having nearly the same wave length (about 100 km) as in Fig. 3.20(b), in constrat to $exp\_galchen = 2.6$ that leads to a monotone ascending ray path. The reason is that the latter one generates a coarse grid spacing which smooths out the gradient of refractive index in the lower atmosphere and renders the curvature (towards the Earth's surface) of the ray path less due to Eq. (3.45). For $\varepsilon_0 = 0.5°$, the slope is already large enough so that the beam penetrates the ducting layer despite of finer resolution $exp\_galchen = 3.6$.

3.5.6. Summary and Discussion

In this section, we assessed the performance of three radar beam tracing methods, 43ERM (well-known and based on atmospheric standard conditions), TORE (partly known from literature and based on actual vertical profiles of refractive index n) and SODE (new method, based on actual profiles of n and introduced in Subsection 3.5.4 by several sensitivity experiments. Both TORE and SODE methods employ actual n data and are rigorously based on Geometrical Optics and its fundamental Fermat's principle. Whereas SODE involves the solution of an initial value problem of an ordinary differential equation, TORE is based on the conservation of an integral quantity of this differential equation along the ray path, known as Snell's law for continuously stratified media. It is documented that 43ERM may expose errors under ducting situations and TORE has to employ an ad hoc approach to allow for the sign switch under standard conditions.

Because 43ERM does not take into account the true environmental conditions at a specific time, it tends to overestimate the beam heights in the case of superrefraction or ducting, especially for low elevations ($\varepsilon_0 \lesssim 1.0°$). However, for conditions which are near-normal in the lowest 1000 m or so of the atmosphere or for higher antenna elevations, it generally works well, as also noted in earlier studies (e.g., Doviak and Zrnic (1993)) and is a commonly-used method among radar specialists today. Such conditions are prevailing for the vast majority of cases, because most radar data are taken from elevations $> 1°$, and superrefraction or ducting conditions occur only occasionally.

When it comes, however, to superrefraction or ducting conditions connected with low elevations, TORE is able to grasp the ducting effect on propagation, where the ad hoc approach plays a key role, without which the beam gets "stuck" at a local elevation $\varepsilon \lesssim 0°$ and propagates purely horizontally further out under more or less standard conditions. Nevertheless, this ad hoc approach is based on very simple linear extrapolation, and its stability and robustness in practice need to be further examined.

Instead, reflections occur naturally with SODE, which means that no special measures are necessary to correctly treat reflection points, and it can also provides reasonable and robust results in all presented tests. Moreover, SODE is more convenient in implementation of the radar forward operator due to its r dependency instead of s. Regarding these facts, we consider SODE as a reference method.

It can be said that if consideration of the actual atmospheric conditions is important, the SODE method is more reliable than TORE from the current view, because there are no ad hoc criteria involved in SODE. However, prerequisite for a successful detailed beam propagation computation is a very good knowledge of the 3D atmospheric state, i.e., n and its vertical slope, which may vary also horizontally (this last point is not taken into account in the present paper). With today's aerological network (sparse number of stations and sparse observation times), this is certainly not the case in general, and the results can only be as good as the input data. However, if one day better, i.e., spatially and temporally more dense, observations should be available, then SODE can play out its advantages. We also see that the vertical resolution is important for the accuracy of beam propagation simulation. In this sense, the vertical resolution in (operational) NWP models today might be not be sufficiently high, but for the future it is foreseeable that much efforts will go into higher model resolution (at least in research), so that then, SODE could be the method of choice.

3.6. Beam broadening and shielding

3.6.1. Beam broadening

For a radar of DWD with azimuthal resolution of 1°, the areal size of one bin ranges from around $0.05\,\mathrm{km}^2$ very close to the radar site to around $3.75\,\mathrm{km}^2$ near the end of the radar effective range, that is a size difference of factor 75. Thus, radar observations are of lower resolution at farther distance, which causes differences in appearance of radar returns close to and far from the radar site. The wider the beam is, the greater the likelihood of sampling a mixture of precipitation types becomes, especially in the vertical considering that ice particles melt and change their shape and composition as they fall. Therefore, it is more realistic to account for beam broadening than evaluate reflectivity solely along the beam axis when using radar data as well as developing the operator.

This is done in our operator, e.g., in Eq. (3.36), by pulse-volume averaging over azimuthal and elevational directions. Numerically, integration is approximated by a sum of values at finite integration points within the integration interval, which means that we have to simulate/evaluate not just the ray path of the beam axis but also those of some auxiliary axes. In our operator the two dimensional Gauss-Legendre quadrature (see Appendix B.1) is implemented, with selectable numbers nh and nv of integration points in azimuthal and elevational directions (see Fig. 3.23). For each integration we have to first estimate the integration intervals $\Delta\alpha$ and $\Delta\varepsilon$. Note that $\Delta\alpha$ expands by factor c' due to the angular averaging, which is given by Blahak (2008a)

$$c' = \frac{\alpha_{3,eff,0} + (\cos\varepsilon_0 - 1)\Delta\alpha[1 - \exp(-1.5\Delta\alpha/\theta_3)]}{\theta_3} \quad . \tag{3.74}$$

Additionally, we scale up $\Delta\alpha$ and $\Delta\varepsilon$ by factor c'' (≈ 1.29) to contain 90% power. Therefore, the actual integration intervals in the operator are $\Delta\alpha = [\alpha_0 - c'c''\phi_3/2, \alpha_0 + c'c''\phi_3/2]$ and $\Delta\varepsilon = [\varepsilon_0 - c''\theta_3/2, \varepsilon_0 + c''\theta_3/2]$.

Next, one selects the number of integration points and calculate their positions. Note that since the integration points are symmetrically distributed around the center of the integration interval, odd numbers are suggested to make sure that the beam axis is among the integration points. For instance, nh and nv are the numbers of integration points for intervals $\Delta\alpha$ and $\Delta\varepsilon$, respectively, the positions of integration points $\{\alpha_0^{ih} : ih = 1, \cdots, nh\}$ and $\{\varepsilon_0^{iv}, iv = 1, \cdots, nv\}$ are determined by formula of Gauss-Legendre quadrature. For each integration point an auxiliary axis has to be simulated and the pulse-volume averaging of a certain bin is done by summing up the values (with corresponding Gauss-Legendre weights) on these auxiliary axes within that bin. Thus, $(nh + nv - 1)$ auxiliary rays - instead of solely one beam axis - are used to represent one single beam.

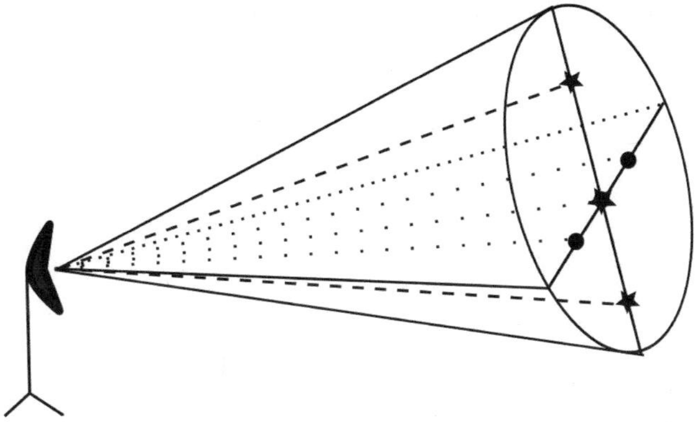

Fig. 3.23.: Sketch of areal integration (plane surrounded by the solid line) with auxiliary interpolation point for the Gauss-Legendre quadrature. The horizontal and vertical axes are integration intervals $\Delta\alpha$ and $\Delta\varepsilon$, respectively, with auxiliary interpolation points, denoted with $\star$ and $\bullet$.

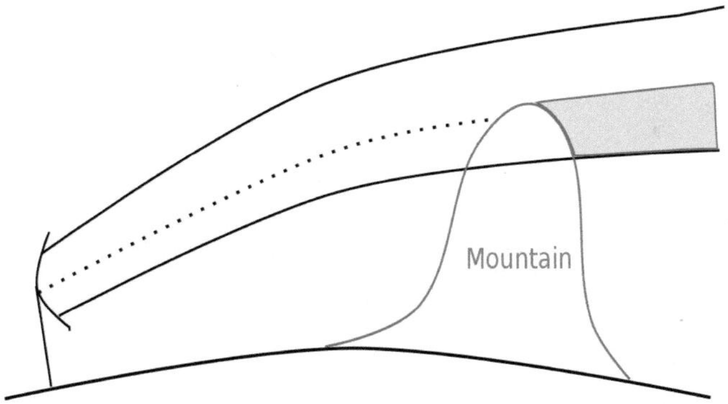

Fig. 3.24.: Beam shielding: the lower portion of the beam strikes the mountain and hence the grey area can not be seen by the radar. The dotted line represents, for instance, an auxiliary ray which is blocked by the mountain and is no taken into account in pulse volume averaging.

3.6.2. Beam shielding

When radars scan in an environment with natural or man-made obstacles, the beams of lowest elevation run very likely into surrounding obstacles. If merely the beam axis is used to represent the beam in simulation, once it hits the obstacle, the beam is totally blocked. But in the reality, a portion of the beam may be still able to travel above the top (or side) of the obstacle, so that the radar can still detect the precipitation behind it (or near it). With pulse-volume averaging this shielding effect can be effectively approximated. Once an auxiliary ray gets blocked somewhere on the way, it and only it will be blocked (see Fig. 3.24). In light of different behaviors of measurements, different averaging strategies are applied to reflectivity and Doppler velocity. For the former one we average over the whole bin, setting the reflectivity in blocked part equal to $0\,\mathrm{mm}^6/\mathrm{m}^3$ with full weight, while for the latter one we discard the blocked part and just average over the unblocked area. This treatment is consistent with the real behaviour of radar measurements.

3.6.3. Sensitivity experiments

Now we are interested in determining the sufficient number of integration points. Loosely speaking[4], the more integration points are, the more accurate are the integrals. But more points potentially bring about more computational time and memory usage. A good choice could be very circumstantial, it depends primarily on the model resolution. For instance, it is advisable to do more averaging when the model resolution is higher than that of radars. Second, it depends on the variability of model states, e.g., larger wind shear needs more averaging for Doppler velocity. On the subject of beam shielding, it is also related to physical properties of obstacles (e.g., height and position relative to radar).

Next, we intend to find an appropriate number of integration points for the given model resolution, thermodynamic profiles and orographic obstacle by a series of experiments in an idealized framework. The first experiment $\mathsf{E}_{1\times1}$ is done without averaging (the first number in subscript refers to nh and the second one to nv) and in each subsequent experiment we increase the number of integration points by two (note:

[4]To be precise, an n-point Gaussian quadrature rule yields already an exact result for polynomials of degree $2n-1$, which means the accuracy of an numerical integration can be limitedly enhanced by increasing integration points.

first in *nv* and then in *nh*). The model setup is same for all experiments: the model grid comprises of $201 \times 100 \times 64$ grid points, with horizontal grid spacing of (0.025°, 0.025°) (approx. 2.8 km). The vertical coordinate is the Gal-Chen coordinate with $exp\_galchen = 2.6$, which generates a moderately increasing grid distance with height (see Tab. 3.2). The thermodynamic profiles are specified analytically and periodic conditions are used at the lateral boundaries. A three dimensional Gaussian-shaped mountain centered at model grid point $(i, j) = (72, 72)$ is set with height of 1000 m and radii of $(75000, 20000)$ m, extending to the northeast corner of the observed area.

A convective system is triggered by three ellipsoidal warm air bubbles centered at $(i, j) = (22.5, 45.5)$, $(i, j) = (26.5, 50.5)$ and $(i, j) = (25.5, 55.5)$, with the same height of 1400 m and three dimensional radii of $(16000, 16000, 1600)$ m (i.e., the bubbles spread out to the surface), within environmental conditions similar to those given by Weisman and Klemp (1982). The maximum potential temperature deviation amounts to 3 K. Because the bubbles are warmer than the surrounding air, they are buoyant and rise freely. As they ascend, they cools at the dry adiabatic rate and the dew point falls, but not as rapidly. So the temperature of bubbles and dew points approach each other and relative humidity of bubbles increases. As the bubbles have cooled down to the dew point, condensation begins, and clouds form. The condensing water releases latent heat energy, which promotes the air lift. During this process, large quantities of water emerge which can cause showers or even thunderstorms.

After about three hours, a large squallline type system has developed and reached mountain area. Fig. 3.25 illustrates the simulated PPI scans at elevation 0.5° for $\mathsf{E}_{1\times1}$, $\mathsf{E}_{1\times3}$ and $\mathsf{E}_{5\times9}$. As shown in Figs. 3.25a and 3.25b, the radar beams are hindered by the mountain in the northeast part and there are no data behind it. In $\mathsf{E}_{1\times3}$, applying pulse-volume averaging, the elevation of the upper auxiliary axis is already large enough to make it penetrate the top of mountain and consequently enlarge the areal cover of reflectivity and Doppler velocity (see Figs. 3.25c and 3.25d, note that Doppler velocities exist even where no reflectivity is present. This is because weighting by reflectivity was not used here, and we will come back to this points later on in the the next chapter). Meanwhile, we can see value cliffs of reflectivity and Doppler velocity on the edge of the mountain, but by employing more and more integration points, this discontinuity can be gradually smoothed away (see Figs. 3.25e and 3.25f).

To gain a deeper insight into effects of increasing integration points, we calculated differences of every two successive experiments and denote them as $\mathsf{E}_{ih\times(iv+2)} - \mathsf{E}_{ih\times iv}$ or $\mathsf{E}_{(ih+2)\times iv} - \mathsf{E}_{ih\times iv}$ (the substration is binwise). The results are shown in Fig. 3.26 for reflectivity and Fig. 3.27 for Doppler velocity. The bins, where at least one of the experiments does not have numerical values (i.e., bins are either blocked due to the orography or outside the model domain), are marked in black color[5]. In Fig. 3.26a, we confirm the occurrence of total beam blockage in $\mathsf{E}_{1\times1}$ in black area. The dark red area arises from comparing small reflectivities of small values ($\leq$ -5 dBZ) with no reflectivity, represented by -99.99 dBZ. From Fig. 3.26b to Fig. 3.26d we can see the impacts of vertical averaging, which continually decrease as integration points increase due to the edge effect. As expected, the horizontal averaging does not exhibit strong impacts because of the relative coarse horizontal resolution. The dark red spots in Fig. 3.26e basically outline the border from reflectivities of small values ($\leq$ -5 dBZ) to no reflectivity, and emerge clearly due to the same reason just described above. These are, however, already greatly reduced in $\mathsf{E}_{5\times9}$, as shown in Fig. 3.26f. For Doppler velocity (see Fig. 3.27), the differences also tendentially fade out with increasing integration points. Meanwhile, a clear stripe structure can be seen which becomes finer as more integration points are employed. The reason is illustrated in Fig. 3.28, where we compare integration schemes with three and five points, for instance. At position P1 the lower outermost point of 5-point integration scheme is under the orography (i.e., the corresponding auxiliary axis is blocked), so its value will not contribute to the integration according to the averaging strategy for Doppler velocity and the other four points build the integration. By doing this, we neglect the point with the smallest value since the wind speed rises with the height, which results in a faster Doppler velocity than the 3-point integration scheme using all three points in this case. When the radar scan moves horizontally closer to mountain, say at position P2, the lower outermost point of 3-point integration scheme encounters the orography as well, analogously, this enhances the value of integral, but to a even higher extent and results in a faster Doppler velocity than 5-point integration at this position. This alternating value relationship brings about the stripe structure. If we compare two integration

[5]In this thesis, we distinguish places where no reflectivity exists and where beams are blocked or outside the model domain. The first one is assigned to value of -99.99 dBZ, while the latter one is given a nonnumerical value.

schemes with even more integration points, this kind of alternation will occur more frequently and we will see more stripes with smaller widths. Figs. 3.27e-3.27f show that the horizontal averaging is negligible here, which is attributable to the model resolution and homogeneous wind profile.

3.6.4. Summary

To conclude, the choice of $(nh,nv) = (5,9)$ already provides hardly improvable results with respect to pulse-volume averaging. Concerning the computational efforts, we can not arbitrarily increase the number of integration points. One must be aware that a strong convective scenario is handled here, representing large inhomogeneity of reflectivity and Doppler velocity. In case of a stratiform rain, probably less integration points can be considered. Therefore, in the sensitive experiments below, the choice of $(nh,nv) = (5,9)$ is adopted to guarantee good averaging results.

3.7. Simulation of attenuated reflectivity

The simulation of attenuated reflectivity corresponds to the hydrometeor and the thermodynamic values including rain water-, graupel/hail- and snow content, air density and air temperature on model grids and follow the so-called graupel scheme (Reinhardt and Seifert, 2006) that considers mass densities $\{q^c, q^i, q^r, q^s, q^g\}$ of cloud water, cloud ice, rain, snow and graupel as prognostic variables in case of the one moment scheme and hail as an additional type of precipitation particle in case of two-moment scheme (Seifert and Beheng, 2006). The simulation code used here is provided by Blahak (2007), which computes the reflectivity Z_e and attenuation coefficient Λ on model grid points, based on the full Mie-scattering scheme and temperature dependent refractive index of the particles. Alternatively, Z_e can be calculated by the Rayleigh approximation together with simple formulas for the refractive index (Oguchi, 1983). Special care is given to the description of melting particles. As well-known, the shape of particles are usually different from sphere and the refractive index of the particle material (mixture of ice, water and air) is considerably sensitive to the type and structure of the mixture and the bulk density. Since no absolutely correct theoretical description of the refractive index of such complex mixtures exists, there are many different formulas available. For that, Blahak (2007) chooses three

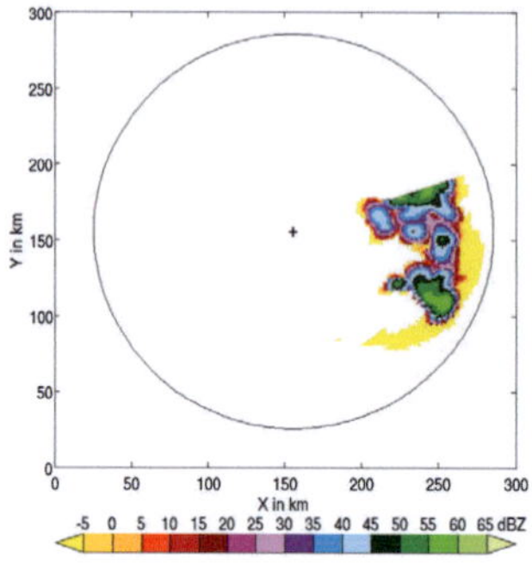

(a) $\mathsf{E}_{1\times1}$

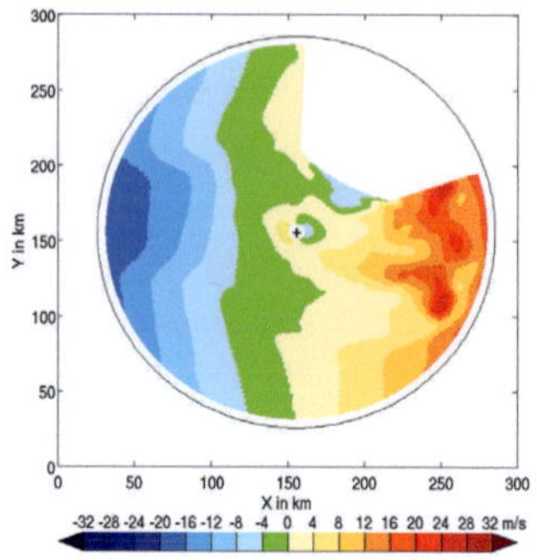

(b) $\mathsf{E}_{1\times1}$

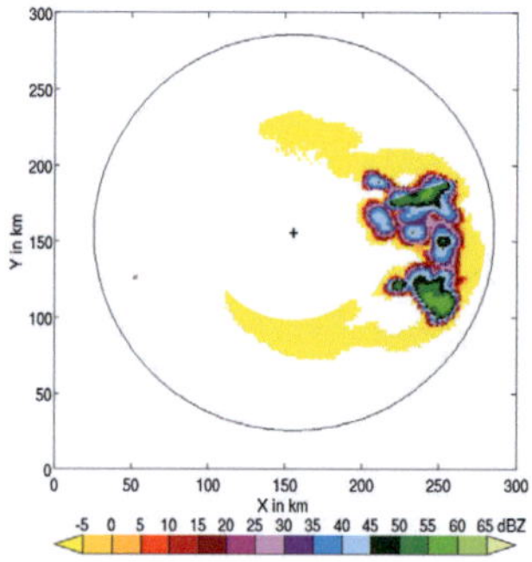

(c) $\mathsf{E}_{1\times3}$

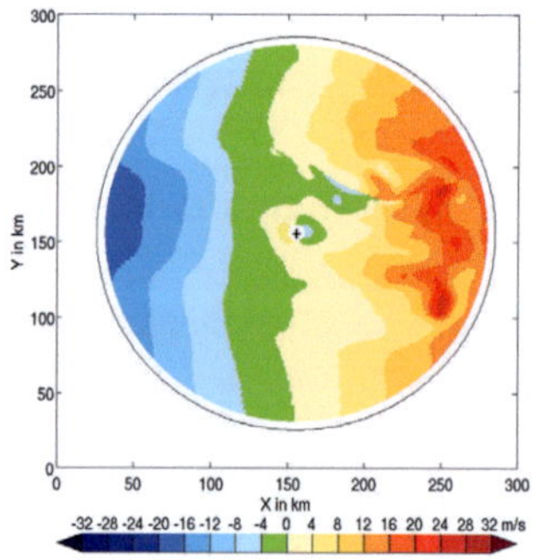

(d) $\mathsf{E}_{1\times3}$

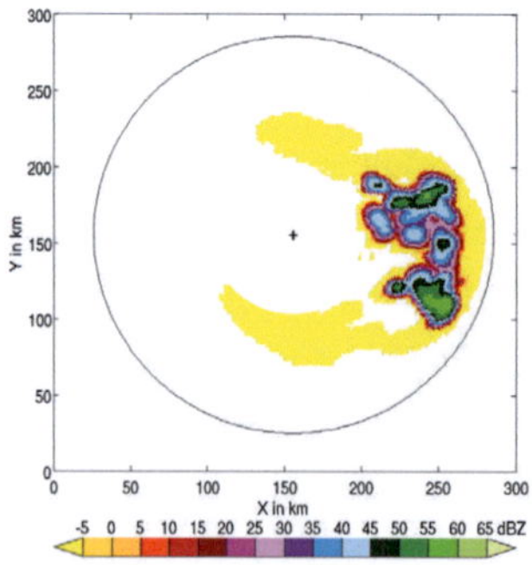

(e) $\mathsf{E}_{5\times9}$

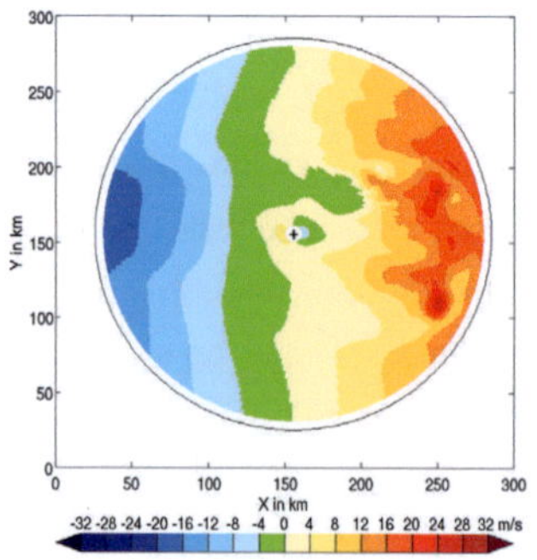

(f) $\mathsf{E}_{5\times9}$

Fig. 3.25.: Sensitivity results at an elevation of 0.5° (PPI mode), based on different numbers of integration points. **Left column**: radar reflectivity in dBZ (see color bar); **Right column**: Doppler velocity in m/s (see color bar)

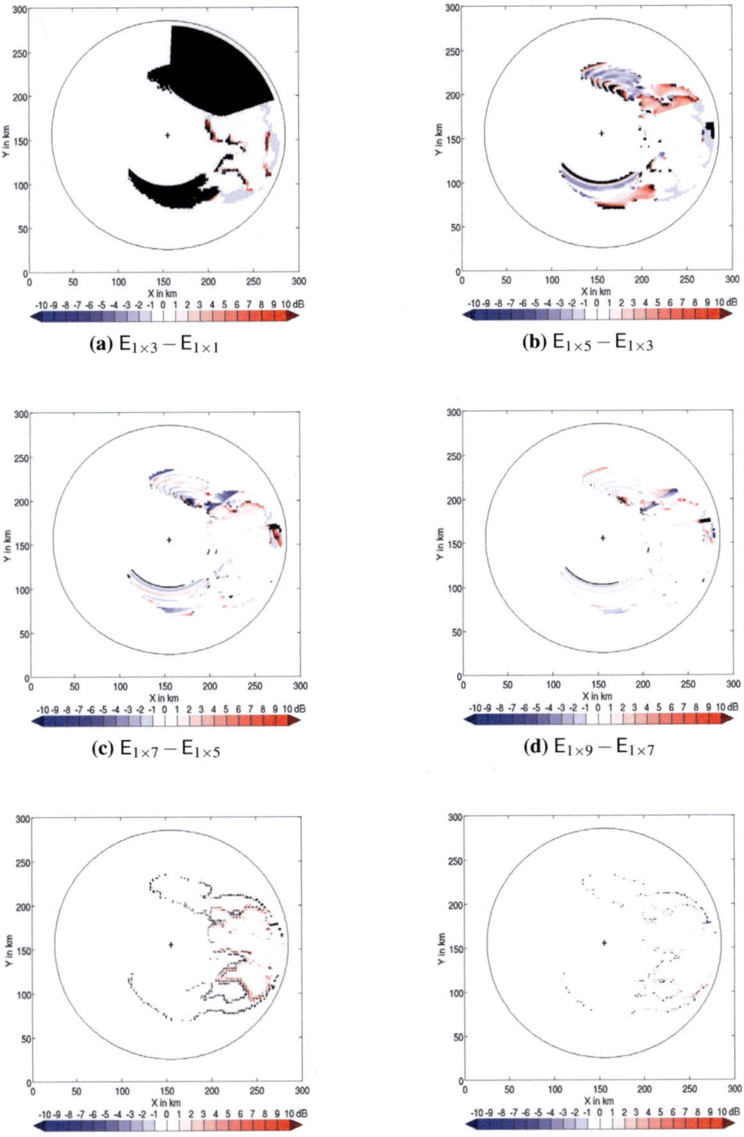

(a) $E_{1\times3} - E_{1\times1}$ **(b)** $E_{1\times5} - E_{1\times3}$

(c) $E_{1\times7} - E_{1\times5}$ **(d)** $E_{1\times9} - E_{1\times7}$

(e) $E_{3\times9} - E_{1\times9}$ **(f)** $E_{5\times9} - E_{3\times9}$

Fig. 3.26.: Reflectivity differences in dB (see color bar) of sensitivity results at an elevation of 0.5° (PPI mode)

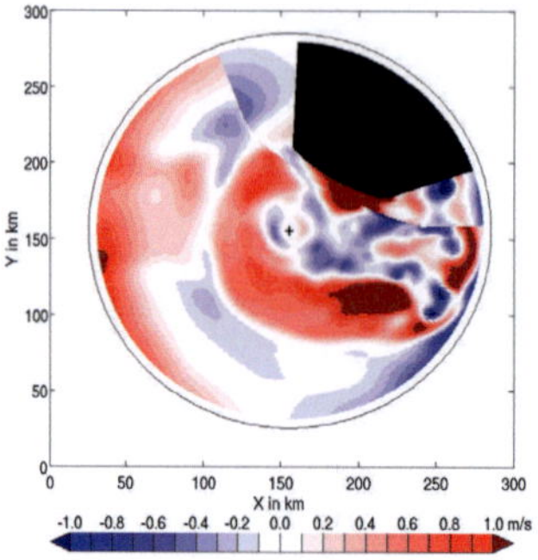

(a) $E_{1\times3} - E_{1\times1}$

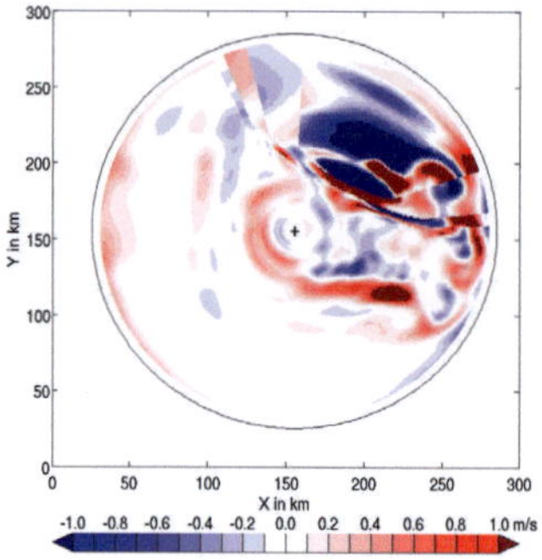

(b) $E_{1\times5} - E_{1\times3}$

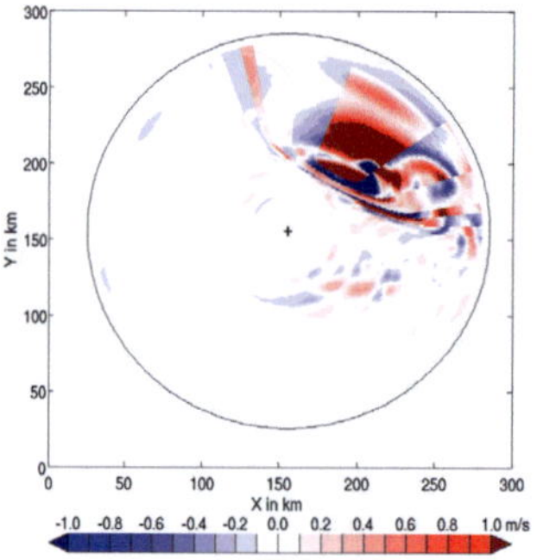

(c) $E_{1\times7} - E_{1\times5}$

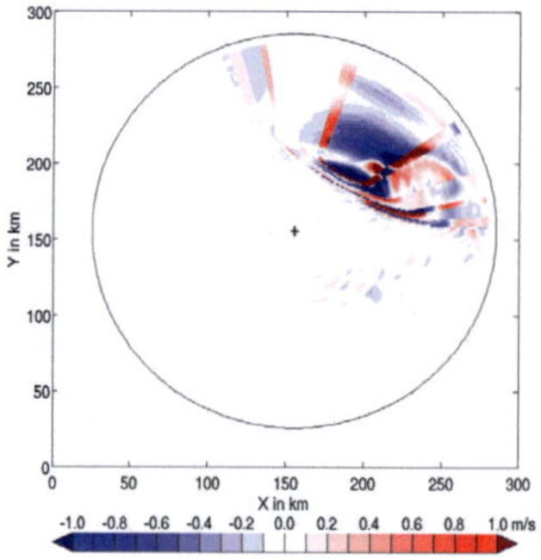

(d) $E_{1\times9} - E_{1\times7}$

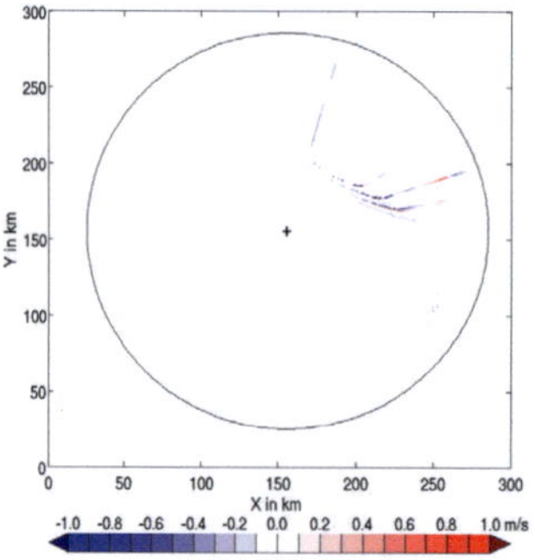

(e) $E_{3\times9} - E_{1\times9}$

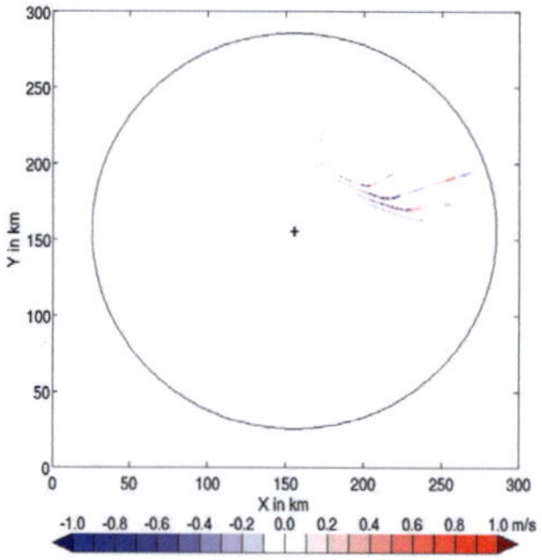

(f) $E_{5\times9} - E_{3\times9}$

Fig. 3.27.: As Fig. 3.26 but for differences of Doppler velocity in m/s (see color bar)

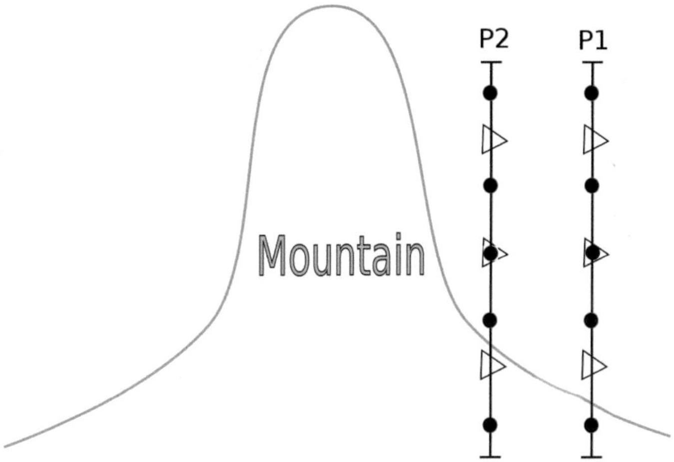

Fig. 3.28.: Vertical section of an azimuthal scan: The straight line represent the integration interval (or beamwidth); 3-point integration scheme: $\triangleright$; 5-point integration scheme: $\bullet$.

popular Effective Medium Approximations (EMA) (Maxwell-Garnett, Wiener and Bruggmann) but considers only spherical particles, where a concentric two-layer-sphere particle model is implemented for e.g., snowflakes as well as melting hail- and graupel particles. Normally, the Mie solution results in lower reflectivities than the Rayleigh approximation for the large particles (except rain drop), but when it comes to the melting particles, the Mie-scattering scheme usually produces higher reflectivities because of the special effective refractive index generated by EMA engenders stronger echoes than the Ogachi's formula. In this connection, more can be seen in Chapter 5.

As noticed in Eq. (3.19), PSDs of the different hydrometeors are required for computation of η_e (or Z_e). Within the operator, PSDs are derived from the prognostic specific mass fractions q^x in a model-consistent way, i.e., using the same assumptions for PSDs (e.g., generalized gamma distribution) and the mass-size- and fallspeed-size-relations (power laws) as the COSMO-model does, and then σ_b can be integrated over PSDs and summed up over all species to get Z_e. We take here graupel as example for a brief derivation of a PSD and for the other hydrometeors we refer to Doms and Schättler (2002).

An expontenial size distribution is assumed for graupel:

$$\mathcal{N}^g(D) = \mathcal{N}_0^g(D)\exp(-\lambda^g D) \quad , \tag{3.75}$$

where $\mathcal{N}_0^g(D) = 4 \times 10^6\,\mathrm{m}^{-4}$ (Rutlege and Hobbs, 1984). The properties aof single graupel particles in the form of power laws are based on Heymsfield and Kajikawa (1986) for their lump graupel ($\rho^g \approx 0.2\ \mathrm{g/cm^3}$) and it is assumed that $m^g = a_m^g D^{3.1}$ with $a_m^g = 169.6$ for the masse-size relation and for the terminal fall velocity: $w_T^g(D) = w_0^g D^{0.89}$ with $w_0^g = 442.0$ (all in the corresponding SI units).

Because it holds for the graupel mass density

$$q^g = \int_0^\infty m^g(D)\mathcal{N}^g(D)dD \quad , \tag{3.76}$$

we replace $\mathcal{N}^g(D)$ in Eq. (3.76) with the right-hand side of Eq. (3.75) and express λ^g with known quantities

$$\lambda^g = f^g(\mathcal{N}_0^g, q^g) \quad , \tag{3.77}$$

and applying Eq. (3.77) in Eq. (3.75), we can obtain $\mathcal{N}^g$.

The extension of the original code to include attenuation coefficient Λ has been done by Jerger und Blahak (Jerger et al., 2012). Meanwhile, lookup-tables have been established for each specie of hydrometeor to avoid the excessive time consumption of the Mie solution. The values vary with water content, temperature (due to the refractive index) and average water content in the ice-water-air mixture. Particularly, each combination of different models for particles with different melting degree (single-/two-shell configurations combined with either internally accumulated melt water or absorption of water in the porous structure of ice) needs to be handled separately. The Mie solution can then be comfortably estimated by interpolation of values in look-up tables. These three approaches (Mie-scattering scheme, lookup tables and Rayleigh approximation) form the modular options for simulation of reflectivity in the operator.

Within the operator, Z_e and Λ are first computed on the COSMO-model grid points and then linearly interpolated to the polar radar grid to subsequently perform Eq. (3.36), which serves as a benchmark to calculate Z_e in this work. For the sake of computational efficiency, we often assume that the beam broadening effect is insignificant and hence pulse-volume averaging can be switched off, which means that reflectivity is just evaluated at the centre of each bin, Eq. (3.36) becomes

$$\langle Z_e^{(R)} \rangle(\vec{r}_0) = Z_e(\vec{r}_0)\ell^{-2}(\vec{r}_0) \quad , \tag{3.78}$$

and the assumption that attenuation is negligible yields

$$\langle Z_e^{(R)} \rangle(\vec{r}_0) = \frac{\int\limits_{\alpha_*-\pi}^{\alpha_*+\pi} \int\limits_{\varepsilon_0-\pi/2}^{\varepsilon_0+\pi/2} Z_e(r_0,\alpha,\varepsilon) f_e^4(\alpha,\varepsilon) \cos\varepsilon d\varepsilon d\alpha}{\int\limits_{\alpha_*-\pi}^{\alpha_*+\pi} \int\limits_{\varepsilon_0-\pi/2}^{\varepsilon_0+\pi/2} f_e^4(\alpha,\varepsilon) \cos\varepsilon d\varepsilon d\alpha} \quad . \tag{3.79}$$

3.8. Simulation of Doppler velocity

In this work, we aim to devise a comprehensive and flexible simulator of Doppler velocity which accounts for the reflectivity weighting for wind velocity and fall velocity of hydrometeors as well as effects like beam bending and broadening.

The 3D-wind vector is projected on the slanted direction of the radar beam and the model counterpart of the measured Doppler velocity is given by Lindskog et al. (2004)

$$v_r = (u \sin\alpha + v \cos\alpha) \cos\varepsilon + (w - \overline{w}_t) \sin\varepsilon \quad , \tag{3.80}$$

where α is the radar antenna azimuth and ε is local elevation. ε can be estimated by the online methods for beam propagation introduced in Section 3.5 (i.e., TORE snd SODE). u, v and w are the zonal, meridional and vertical components of the wind vector from the model, respectively. The meteorological convention for winds is that u component is positive for a west to east flow (west wind), the v component is positive for a south to north flow (south wind) and w is positive for a upward flow. $\overline{w}_t$ is the average terminal fall speed of hydrometeors, which is defined by:

$$\overline{w}_t := \begin{cases} \left(\frac{\rho_0}{\rho}\right)^{0.5} \frac{\sum\limits_{k \in S} \int_0^\infty \sigma_{b_k}(D_k) w_{t_k}(D_k) \mathcal{N}^k(D_k) dD_k}{\eta}, & \text{if weighting by reflectivity;} \\ \left(\frac{\rho_0}{\rho}\right)^{0.5} \frac{\sum\limits_{k \in S} \int_0^\infty w_{t_k}(D_k) \mathcal{N}^k(D_k) dD_k}{\sum\limits_{k \in S} \int_0^\infty \mathcal{N}^k(D_k) dD_k}, & \text{otherwise,} \end{cases} \tag{3.81}$$

where ρ is the air density, ρ_0 is the reference (mostly surface air) density and $w_{t_k}(D_k)$ is the terminal fall velocity of a single hydrometeor as function of diameter D_k. Within the COSMO-model, w_{t_k} is computed by the formula $w_{t_k}(D_k) = aD_k^b$, where a and b are different for each hydrometeor class k. The consideration of terminal fall velocity is important, especially for assimilation of high-elevation radar data. Currently, we use some constant value for $\overline{w}_t$ (e.g., 5 m/s) because its implementation with full model-consistent coupling to $\mathcal{N}^k(D_k)$, $\sigma_{b_k}(D_k)$ and $w_{t_k}(D_k)$ could not be accomplished during the course of this work but it will be done in the near future.

If one applies 43ERM to approximate the radar beam, Eq. (3.80) becomes (Järvinen et al., 2009)

$$v_r = (u \sin\alpha + v \cos\alpha)\cos(\varepsilon + \delta\varepsilon) + (w - \overline{w}_t)\sin(\varepsilon + \delta\varepsilon) \quad , \tag{3.82}$$

where

$$\delta\varepsilon = \arctan\left(\frac{r\cos\varepsilon}{r\sin\varepsilon + \frac{4}{3}R_E + h_0}\right) \tag{3.83}$$

takes approximately the curvature of the earth into account.

Taking into account beam broadening, Eq. (3.40) is applied in (almost) full detail (no range weighting). By neglecting the effect of beam broadening, Eq. (3.40) can be reduced to:

$$\langle v_r^{(R)} \rangle(\vec{r}_0) = \vec{v}(\vec{r}_0) \cdot \vec{e}_r - (\vec{e}_3 \cdot \vec{e}_r)\overline{w}_t(\vec{r}_0) \quad . \tag{3.84}$$

Under the assumption that the hydrometeor fall speed is negligible, Eq. (3.40) can be rewritten as:

$$\langle v_r^{(R)} \rangle (\vec{r}_0) = \frac{\int\limits_{\alpha_*-\pi}^{\alpha_*+\pi} \int\limits_{\varepsilon_0-\pi/2}^{\varepsilon_0+\pi/2} (\vec{v}(r_0,\alpha,\varepsilon)\cdot\vec{e}_r) \frac{\eta(r_0,\alpha,\varepsilon)}{\ell^2(r_0,\alpha,\varepsilon)} f_e^4(\alpha,\varepsilon) \cos\varepsilon d\varepsilon d\alpha}{\int\limits_{\alpha_*-\pi}^{\alpha_*+\pi} \int\limits_{\varepsilon_0-\pi/2}^{\varepsilon_0+\pi/2} \frac{\eta(r_0,\alpha,\varepsilon)}{\ell^2(r_0,\alpha,\varepsilon)} f_e^4(\alpha,\varepsilon) \cos\varepsilon d\varepsilon d\alpha} . \quad (3.85)$$

Another simplification can be achieved if we neglect the weighting by (attenuated) reflectivity:

$$\langle v_r^{(R)} \rangle (\vec{r}_0) = \frac{\int\limits_{\alpha_*-\pi}^{\alpha_*+\pi} \int\limits_{\varepsilon_0-\pi/2}^{\varepsilon_0+\pi/2} (\vec{v}(r_0,\alpha,\varepsilon)\cdot\vec{e}_r) f_e^4(\alpha,\varepsilon) \cos\varepsilon d\varepsilon d\alpha}{\int\limits_{\alpha_*-\pi}^{\alpha_*+\pi} \int\limits_{\varepsilon_0-\pi/2}^{\varepsilon_0\pi/2} f_e^4(\alpha,\varepsilon) \cos\varepsilon d\varepsilon d\alpha} - \frac{\int\limits_{\alpha_*-\pi}^{\alpha_*+\pi} \int\limits_{\varepsilon_0-\pi/2}^{\varepsilon_0+\pi/2} (\vec{e}_3\cdot\vec{e}_r)\overline{w}_t(r_0,\alpha,\varepsilon) f_e^4(\alpha,\varepsilon) \cos\varepsilon d\varepsilon d\alpha}{\int\limits_{\alpha_*-\pi}^{\alpha_*+\pi} \int\limits_{\varepsilon_0-\pi/2}^{\varepsilon_0+\pi/2} f_e^4(\alpha,\varepsilon) \cos\varepsilon d\varepsilon d\alpha} . \quad (3.86)$$

At this point we would like to summarize all available simplifications. With respect to beam bending, we can choose among the simple offline method 43ERM and sophisticated online methods TORE and SODE. In terms of beam broadening, we can take this effect into account by integration or neglect it. The computation of reflectivity can be done either by Mie solution or by Rayleigh approximation, with or without consideration of attenuation. We can also decide to apply weighting by reflectivity or not in computation of Doppler velocity. Those simplifications can save computational efforts but might lead to accuracy loss in simulations, and their significances will be evaluated in Chapter 5. Before that, we will first give a brief overview of the operator from the implementation point of view in the next section.

4. Programming aspects of the radar forward operator

With respect to the program design, applicability and efficiency of the operator code on vector-parallel supercomputers is a major concern. Currently, DWD operates two independent NEC SX-9 clusters, one for the operational weather forecast, the other one for research and development. In this section, we explain the computer implementation of the operator for the NEC SX-9 clusters of DWD. Because the operator is implemented in the framework of the COSMO-model, it shares the same programming language, namely FORTRAN 90 or 95. Recall that another goal of the code design stated in the last chapter is to have a flexible modular operator that offers different options for each module. The control flags introduced in Tab. 4.1 are used in implementation to switch on/off modules or to select specific options, and they can be specified via a Fortran 90 namelist file.

Logical flag (default = .true.)	Function
lout_geom	Output of heights h and elevations ε of radar bins
loutradwind	Output of Doppler velocity v_r
loutdbz	Output of reflectivity Z_e
lfall	Taking fallspeed w_t into account
lweightdbz	Taking weighting by reflectivity into account
lextdbz	Taking attenuation into account
lsmooth	Taking pulse averaging into account
lonline	Simulation of beam propagation with SODE
lfirst_cmp	Indication of the first call of the operator in a model run

Tab. 4.1.: Function of the control flags in algorithms

4.1. Vectorization

Each cluster of DWD has 14 nodes with 4096 GB/s shared memory bandwidth per node and $2 \times 128\,\mathrm{GB/s}$ bidirectional internode bandwidth. There is 512 GB physical memory per node, the complete system having 7 TB physical memory. Each node has 16 processors. The NEC SX-9 processors run at 3.2 GHz, with eight-way replicated vector pipes, each having two multiply units and two addition units; this results in a peak node performance of 102.4 GFlops/s (= 102,4 billion operations per second) and $102.4\,\mathrm{GFlops} \times 14 = 22.93\,\mathrm{TFlops/s}$ peak system performance. For non-vectorized code, there is a scalar processor that runs at half the speed of the vector unit, i.e., 1.6 GHz. This gives a hint that vectorization[1] of code architecture can accelerate the model runs and save enormous computational time, which is especially of great importance for the operational use. Vectorization entails changes in the order of operations within a loop, so vectorization is only possible if this change of order does not affect the calculation results, which means no data dependency between loop iterations exists.

The NEC SX-9 processors are able to vectorize the innermost loop, so we should make the innermost loop as long as possible in the operator implementation. Naively, we can deal with all radar bins of a radar station in three loops over *naz* (number of azimuths), *nra* (number of range) and *nel* (number of elevations), thereby the length of the innermost loop can just be maximum among them. However, under the assumption of a static radar grid geometry (i.e, using 43ERM), each radar bin is independent from the others and hence we can build one single vectorizable loop over $nrp = naz \cdot nra \cdot nel$ of all bins. The formulas:

$$irp = iaz + (ira - 1) \cdot naz + (iel - 1) \cdot naz \cdot nra \quad , \tag{4.1}$$

$$iaz = MOD(np - 1, naz) + 1 \quad , \tag{4.2}$$

$$ira = MOD(np - 1/naz, nra) + 1 \quad , \tag{4.3}$$

$$iel = (np - 1)/(naz \cdot nra) + 1 \quad , \tag{4.4}$$

[1] In Computer science the process of converting an algorithm from a scalar implementation, which does an operation on one pair of operands at a time, to a vector process where a single instruction can refer to a vector (series of adjacent values) is called vectorization (Piper, 2012).

allow for a unique bidirectional mapping between *irp* and (ira, iaz, iel), where *irp* is the numeration index of the single loop and *ira*, *iaz* and *iel* are indices of range, azimuths and elevations, respectively. In this way, we maximize the length of the innermost loop and from the loop index *irp* we can easily estimate azimuth, range and elevation of a specific bin. Meanwhile, it also reduce the communication overhead, since only one communication step is required instead of three. In case of the dynamical radar grid geometry (i.e., using TORE or SODE), however, this kind of vectorization is not totally feasible due to the dependency on the radial direction. Therefore, the code is vectorized over $naz \cdot nel$ as it is later discussed.

4.2. Parallelized code

The idea of parallelization is to distribute the computation efforts to each processor, with an appropriate load balancing, which refers to the practice of distributing work among processors as equally as possible so that all processors are kept busy all of the time. Load balancing is important for decent performance of parallel programs. For example, if all processors are subject to a barrier synchronization point, the slowest processor will determine the overall performance, so load balancing can be considered as a minimization of the processor's idle time. The COSMO-model exploits static coarse-grained parallelism through the use of MPI (Message Passing Interface) tasking (Vetter et al., 1999). The entire COSMO-model domain is split horizontally in a number of regular rectangles with equal base area, so that each processor computes the time integration of the model equations only for such a subdomain and all processors have comparable work to do. In each time step, the exchange of data values across domain borders is required for the finite-difference calculation of horizontal gradients at the domain boundaries, therefore, communication between neighboring processors has to be executed. However, since the communication can be very time-consuming for current supercomputer architectures, we attempt to minimize communication steps as much as possible.

Depending on methods of simulating radar beam propagation, two kinds of parallelization strategies are viable. In case of the time-constant 43ERM, we have to calculate the geographic latitudes, lontitudes and heights $(\varphi_g, \lambda_g, h)$ of the radar bins, using (Blahak, 2004):

$$\varphi_g = \arcsin\left(\sin\varphi_{g0}\cos\left(\frac{s}{R_{eff}+h_0}\right) + \cos\varphi_{g0}\sin\left(\frac{s}{R_{eff}+h_0}\right)\cos\alpha\right) \quad , \tag{4.5}$$

$$\lambda_g = \begin{cases} \lambda_{g0} + \arccos\left(\dfrac{\cos\left(\dfrac{s}{R_{eff}+h_0}\right) - \sin\varphi_{g0}\sin\varphi_g}{\cos\varphi_{g0}\cos\varphi_g}\right); & \text{if } \sin\alpha \leq \pi, \\ \lambda_{g0} - \arccos\left(\dfrac{\cos\left(\dfrac{s}{R_{eff}+h_0}\right) - \sin\varphi_{g0}\sin\varphi_g}{\cos\varphi_{g0}\cos\varphi_g}\right), & \text{if } \sin\alpha > \pi \end{cases} \tag{4.6}$$

and Eqs. (3.53-3.54), where $(\varphi_{g0}, \lambda_{g0})$ is the horizontal position of the radar, and then transform them to the rotated coordinates (φ, λ, h) (cf. Section 2.1). These calculations have to be done only once at the beginning of a model run because of offline characteristic of 43ERM. The parallelization is realized in the framework of the static domain decomposition of the COSMO-model. Each processor computes first geometric coordinates of all possible bins for each radar and then determines the observable bins in its own domain.

Algorithm 1 calc_geometry_43ERM

1: **for** each radar station **do**
2:     **for** *irp* := 1 to *nrp* **do**
3:         Calculate (φ, λ, h) of the bin
4:         Estimate index *imp*, indicating the model grid box, in which the bin is located
5:         Calculate 3D weight $\vec{w}_3$ and determine if the bin is observable
6:         Estimate ε and h and save index *irp* if the bin is observable
7:     **end for**
8:     Estimate the total number *nobs* of observable bins
9: **end for**

In Alg. 1, the index *imp* is used to remember the model grid (i, j, k) and works in a similar way as *irp* in terms of vectorization (cf. Eqs. (4.1-4.4)). In Step 3, $\vec{w}_3$ reflects the distances to the upper southwest corner of imp^{th} model grid box (cf. Fig. 2.5): the farther the distance is, the smaller value $\vec{w}_3$ gains. A failure value[2] is assigned to $\vec{w}_3$ if the bin is outside the processor domain or unobservable. So for each observable bin we save *irp*, *imp* and $\vec{w}_3$ for subsequent calculations, from which we can extract the knowledge of radar polar coordinates of each observable bin, the model grid box containing it and its ratio to model values.

The processor now can calculate fall speed (Alg. 2), Doppler velocity (Alg. 3) and reflectivity (Alg. 4) on those bins and save them in respective vectors. In Algs. 2 and 4, one computes first w_t, Z_e or Λ on model grids and then interpolate them trilinearly onto radar grids, according to *irp*, *imp* and $\vec{w}_3$. Step 3 in Alg. 3 is done by trilinear interpolation as well.

Algorithm 2 calc_mod_fallspeed(time)

```
Calculate fallspeeds w_t on all model grid points within the processor domain
for each radar station do
    for iobs := 1 to nobs do
        Calculate w_t and save
    end for
end for
```

Algorithm 3 calc_mod_radialwind(time)

```
1:  for each radar station do
2:      for iobs := 1 to nobs do
3:          Calculate wind vector (u, v, w)
4:          if lfall then
5:              Calculate v_r and save, using Eq. (3.82) with w_t calculated from Alg. 2
6:          else
7:              Calculate v_r and save, using Eq. (3.82) with w_t = 0.0
8:          end if
9:      end for
10: end for
```

[2]For all quantities in the operator their failure values are set to be -999.99.

Algorithm 4 calc_mod_reflectivity(time)

Calculate reflectivities Z_e on all model grid points within the processor domain
if *lextdbz* **then**
    Estimate attenuations Λ on all model grid points within the processor domain
end if
for each radar station **do**
    for $iobs := 1$ to $nobs$ **do**
        Calculate Z_e and save
        if *lextdbz* **then**
            Calculate Λ and save
        end if
    end for
end for

The last remaining work is to output the data of each radar station in separate files (Alg. 5), thereby the data of h, ε, v_r, Z_e and possibly Λ of a single radar station are collected to a single processor. Since the output files are supposed to be written in radar polar coordinates for the further processing, the indices irp have to be collected as well to provide ranges, azimuths and elevations of radar bins. If taking attenuation into account, Λ (cf. Eq. (3.22)) are summed up along single ray paths to obtain attenuation factor ℓ (cf. Eq. (3.21)) and attenuated reflectivity Z_e at a particular range. The output files are in two formats: ASCII and NETCDF[3]. The ASCII files allow us to plot the results with graphical visualization packages (in this work, Matlab has been used for this purpose). The NETCDF files follow the feedback file definition of DWD, given by Rhodin (2012), and serve as inputs for the data assimilation step. Notice that if there are considerably less radar stations than processors, Alg. 5 can be very imbalanced, but it does restrict the expensive communication to a minimum.

In terms of this parallelization strategy, the bins of a specific radar might be distributed asynchronously over different neighboring processor domains: the bins are much denser distributed for regions close to the radar than for remote regions and the number of bins per processor domain depends on the radar position. This could cause some unavoidable load imbalance.

[3]NetCDF (Network Common Data Form) is a set of software libraries and self-describing, machine-independent data formats that support the creation, access, and sharing of array-oriented scientific data. It is initially developed by the Unidata program at the University Corporation for Atmospheric Research (UCAR) and is widely used in climatology, meteorology and oceanography applications (e.g., weather forecasting, climate change).

Algorithm 5 output_radar

```
1: for each radar station do
2:     Determine the processor x to process data
3:     if lout_geom then
4:         Gather h and ε onto processor x
5:     end if
6:     if loutradwind then
7:         Gather v_r onto processor x
8:     end if
9:     if loutdbz or lweightdbz then
10:        Gather Z_e onto processor x
11:        if lextdbz then
12:            Gather Λ onto processor x
13:        end if
14:    end if
15:    Gather irp onto processor x
16:    for iobs := 1 to nobs do
17:        Estimate range, azimuth and elevation from irp
18:    end for
19:    if lout_geom then
20:        for iobs := 1 to nobs do
21:            According to results of Step 17, re-sort h and ε into hrpolar and epolar
22:        end for
23:        Write out hrpolar and epolar in two binary files, respectively
24:    end if
25:    if loutradwind then
26:        for iobs := 1 to nobs do
27:            Re-sort v_r into vrpolar
28:        end for
29:        Write out vrpolar in a binary file
30:    end if
31:    if loutdbz then
32:        for iobs := 1 to nobs do
33:            Re-sort Z_e into zrpolar
34:        end for
35:        if lextdbz then
36:            for iobs := 1 to nobs do
37:                Re-sort Λ and calculate ℓ and save it in zepolar
38:            end for
39:            Correct zrpolar with zeploar
40:            Write out zepolar in binary file
41:        end if
42:        Write out zrpolar in binary file
43:    end if
44: end for
```

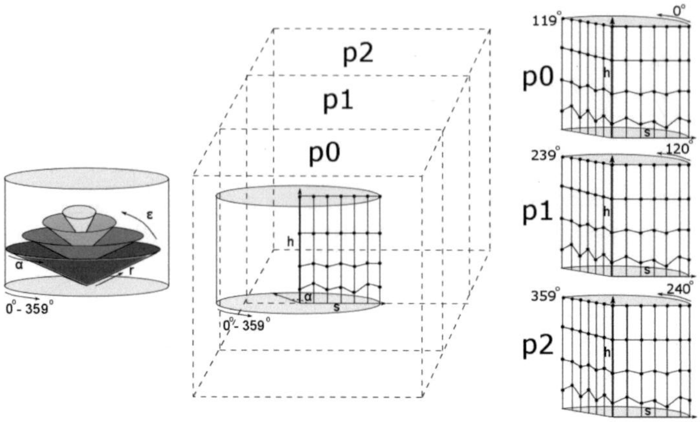

Fig. 4.1.: Sketch of the parallelization concept with an auxiliary grid: **Left:** the radar volume scan is contained in a cylinder; **Middle:** the model domain (dashed line) is, for instance, divided into three processor domains (p0, p1, p2). Discretize the cylinder with an auxiliary grid (α, s, h), see details in text, interpolate the data from the model grid onto the azimuthal slices. **Right:** data along with the azimuthal slices are equally distributed to three processors, i.e., p0 acquires 120 slices from 0° to 119°, p1 from 120° to 239°, p2 from 240° to 359°.

If SODE is used to simulate the beam propagation, geometric coordinates of radar bins must be computed every time step due to temporal variability of refractivity. Since SODE computes iteratively the heights of radar bins along each ray path from the radar site radially outwards, costly and very imbalanced communication steps are expected when ray paths pass through the processor borders because processors have to wait for the results of the others. To avoid this and to ease the organization of communication, an auxiliary grid structure is proposed which consists of azimuthal slices centered around the radar stations (see Fig. 4.1). Each grid point is identified by three elements (α, s, h), where α refers to azimuths of radar scanning and s equidistant arc length along the surface reduced to MSL. The subspace (s, h) represents vertical "slices" at constant azimuths. The auxiliary structure has the same number of vertical levels as the model and h is obtained by horizontal interpolation from heights of four surrounding model grid points. Let nal be the dimension in s and nhl the dimension in h, then there are $ngp = naz \cdot nal \cdot nhl$ auxiliary grid points in total. As the first step, each processor has to determine the valid points, i.e., in its own domain, using Alg. 6.

Algorithm 6 calc_geometry_grid

```
1: for each radar station do
2:     for igrd := 1 to ngp do
3:         Calculate (φ, λ) of the auxiliary grid point
4:         Estimate index imp, indicating the model grid box, in which the point is located
5:         Calculate 2D weight w⃗2
6:         Calculate h for each point by bilinear interpolation, using w⃗2
7:         Save index igrd if the point is valid
8:     end for
9:     Estimate the total number ngrd of valid points
10: end for
```

In Alg. 6, Step 3 includes Eqs. (4.5-4.6) and transformation to rotated coordinates. It is worth mentioning that just two dimensional horizontal weights $\vec{w}_2$ need to be calculated here and a failure value is assigned if the point is outside the processor domain. The index *igrd* is used to save indices of (α, s, h) of each auxiliary grid point and works in a similar way as *irp*.

After geometric coordinates of all valid auxiliary grid points have been computed on a certain processor, the fall speed, Doppler velocity, reflectivity and refractivity can be horizontally interpolated onto those points. In Algs. 7-10, the estimations of quantities on the auxiliary grid points are done by the bilinear interpolation from the corresponding model values, based on $\vec{w}_2$ and *imp*. To solve Eq. (3.68), two quantities n and $\frac{dh}{d\mathrm{n}}$ are desired, but in order to lighten the communication load, a resultant quantity $\frac{1}{\mathrm{n\_grd}} \cdot \frac{\mathrm{n\_low} - \mathrm{n\_up}}{h\_\mathrm{low} - h\_\mathrm{up}}$ in Step 9 is introduced to approximate $\frac{1}{\mathrm{n}} \cdot \frac{d\mathrm{n}}{dh}$.

Algorithm 7 calc_grd_uvw(time)

```
for each radar station do
    for igrd := 1 to ngrd do
        Calculate (u, v, w)
    end for
end for
```

Algorithm 8 calc_grd_fallspeed(time)

Calculate w_t on native model grid points within the processor domain
for each radar station **do**
    for $igrd := 1$ to $ngrd$ **do**
        Estimate w_t_grd
    end for
end for

Algorithm 9 calc_grd_reflectivity(time)

Calculate Z_e on native model grid points within the processor domain
if *lextdbz* **then**
    Calculate Λ on native model grid points within the processor domain
end if
for each radar station **do**
    for $igrd := 1$ to $ngrd$ **do**
        Estimate Z_e_grd
        if *lextdbz* **then**
            Estimate Λ_grd
        end if
    end for
end for

Algorithm 10 calc_grd_rfridx(time)

for each radar station **do**
    for $igrd := 1$ to $ngrd$ **do**
        Estimate T, e and p
        Estimate n_grd using Eq.(3.49)
        Estimate n_low on the lower model level in an analogous manner
        Estimate n_up on the upper model level in an analogous manner
        Estimate h_low on the lower level
        Estimate h_up on the upper level
        Update n_grd using Eq. $\mathsf{n\_grd} = \frac{1}{\mathsf{n\_grd}} \cdot \frac{\mathsf{n\_low} - \mathsf{n\_up}}{h\_\mathrm{low} - h\_\mathrm{up}}$
    end for
end for

After the necessary data have been interpolated to the auxiliary slices, it is time to distribute data among processors and to let each processor do its portion in parallel. In the implementation, the so-called block distribution (Vetter et al., 1999) is applied, where the slices are divided into $nprocs$ parts ($nprocs$ is the number of processors) and the slices in each part is consecutive in terms of the azimuthal index from $n1 = 0$ to $n2 = 359$. For instance (see Fig. 4.1), if three processors are employed to process the 360 slices, processor 0 does slices 0-119, processor 1 slices 120-239, processor 2 slices 240-359. If the total number of slices is not divisible by the number of processors, for example $nprocs = 16$, that is $360 = 16 \times 22 + 8$, we should adjust the way in distributing slices. The idea adopted here is that processors $0, \ldots, 8-1 = 7$ are assigned $22+1 = 23$ slices each and the other processors are assigned 22 slices. The following Alg. 11 depicts how the ranges of slices for all processors are determined.

Algorithm 11 para_range($n1, n2, nprocs, nbl\_az, istart, iend$)

$iwork1 = (n2 - n1 + 1)/nprocs$
$iwork2 = MOD(n2 - n1 + 1, nprocs)$
for $i := 0$ to $nprocs - 1$ **do**
    $istart(irank) = irank \cdot iwork1 + n1 + MIN(irank, iwork2)$
    $iend(irank) = istart(irank) + iwork1 - 1$
    if $iwork2 > irank$ **then**
        $iend(irank) = iend(irank) + 1$
    end if
    $iend(irank) = iend(irank) + 2 \cdot nbl\_az$
end for

Once the range of slices for each processor is estimated, data can be distributed accordingly by Alg. 12.

Algorithm 12 distribute_onlineinfos(x)

1: Estimate the range of slices for each processor using Alg. 11
2: Distribute specific data x to each processor according to results of Step 1

After distribution of all the necessary data, the radar grid geometry, especially the radar bin heights can be computed independently on each processor (Alg. 13). Notice that the previous loop over nrp is here bisected into two loops: one over range index nra, another one over nae, which is joint index over azimuths and elevations. This is done because of the dependency arising from iterative computations in radial direction, which makes vectorization in the dimension of range impossible. In Alg. 13, a flag ensures that the blocked rays are no longer processed once they encounter obstacles. Step 6 is based on Eqs. (3.57) and (3.72); $\vec{w}_2$ gives the appropriate weight for bilinear interpolation within the igf^{th} auxiliary grid field and the failure value is assigned if a bin is unobservable; Step 10 is done by bilinear interpolation from the auxiliary grid, with $\vec{w}_2$, $igrd$ and igf provided.

Algorithm 13 calc_geometry_sode(time)

1: **for** each radar station **do**
2:     **for** $ira := 1$ to nra **do**
3:         **for** $iae := 1$ to nae **do**
4:             **if** $flag(ira, iae) == 1$ **then**
5:                 Solve Eq. (3.68) to obtain $\frac{dh}{dr}$(ira,iae) and h(ira,iae)
6:                 Calculate $el(ira, iae)$ and $s(ira, iae)$
7:                 Estimate index igf of the auxiliary grid field, in which bin is located
8:                 Calculate 2D weight $\vec{w}_2$ and determine if the bin is observable
9:                 Calculate $\mathtt{n}(ira, iae)$ if the bin is observable and set $flag(ira, iae) = 1$
10:             **end if**
11:         **end for**
12:     **end for**
13:     Estimate the total number nobs of observable bins
14: **end for**

Then, we can evaluate the fall speed of hydrometeors and radar quantities on observable bins by Algs. 14-16, where the operations are basically bilinear interpolation from the auxiliary grid.

Finally, the collection of the whole radar station data sets on single output processors and computation of attenuation factors are performed in the same output subroutine as

Algorithm 14 calc_mod_fallspeed_sode(time)

```
for each radar station do
    for iobs := 1 to nobs do
        Calculate w_t
    end for
end for
```

Algorithm 15 calc_mod_radialwind_sode(time)

```
for each radar station do
    for iobs := 1 to nobs do
        Calculate (u, v, w)
        if lfall then
            Calculate v_r using Eq. (3.82) with w_t calculated from Alg. 14
        else
            Calculate v_r using Eq. (3.82) with w_t = 0.0
        end if
    end for
end for
```

Algorithm 16 calc_mod_reflectivity_sode(time)

```
for each radar station do
    for iobs := 1 to nobs do
        Calculate Z_e
        if lextdbz then
            Calculate Λ
        end if
    end for
end for
```

for the time-constant 43ERM (Alg. 5). Obviously, compared to 43ERM, additional communication steps are necessary which are costly but lead to a better balanced computation of beam propagation because each processor has more or less the same number of azimuthal slices to deal with. A main improvement could be, if it would be possible to organize the output of the data to radar station files in a somewhat different way, so that each processor (not only 1 processor per station) could be involved here. However, no good solution has been found until now.

4.3. Organization of the radar forward operator

At last, we summarize the whole program design in Alg. 17. The outermost loop is time-stepping, the operator conducts simulations and writes out results in interval timesteps. As mentioned before, Algs. 1 and 6 are time-invariant and have to be executed only once at the first time step, but Alg. 13 must be done every time as refractivity changes with time.

For the sake of simplicity, only the non-averaging implementation is introduced above. In case of accounting for pulse averaging, the additional auxiliary rays have to be defined for the numerical quadrature (see Section 3.6) and then essentially the same calculations are executed on those rays as for the non-averaging case. A few more differences occur in the output subroutine, where the with averaging coherent integration steps are carried out and observable bins are just determined.

This chapter can also serve as a manuscript for a rudimentary impression of the technical realization of the operator. Of course, there are some other programming details not described here. For example, the subroutine output_obs_dwd, which reads the radar observations from data base of DWD and write observed and simulated values to the above-mentioned NETCDF feedback file, optionally with data thinning (see Section 6.8).

Algorithm 17 organize_radar

```
for time = starttime : timestep : endtime do
    if lonline then
        if lfirst_cmp then
            Call calc_geometry_grid
            for each radar station do
                Call distribute_onlineinfos(hl_grd)
                Call distribute_onlineinfos(nk_grd)
            end for
            lfirst_cmp = .false.
        end if
        if loutradwind or loutdbz then
            Call calc_grd_rfridx(time)
            if loutradwind then
                Call calc_grd_winduvw(time)
            end if
            if loutdbz then
                Call calc_grd_reflectivity(time)
            end if
            for each radar station do
                if loutradwind then
                    Call distribute_onlineinfos(u_grd)
                    Call distribute_onlineinfos(v_grd)
                    Call distribute_onlineinfos(w_grd)
                end if
                if loutdbz then
                    Call distribute_onlineinfos(z_radar_grd)
                    if lextdbz then
                        Call distribute_onlineinfos(z_ext_grd)
                    end if
                end if
                Call distribute_onlineinfos(rfridx_grd)
            end for
            Call calc_geometry_sode(time)
            if loutradwind then
                Call calc_mod_radialwind_sode(time)
            end if
            if loutdbz then
                Call calc_mod_reflectivity_sode(time)
            end if
            Call output_radar
```

Algorithm 17 organize_radar (continued)

```
        else                                                  ▷ 43ERM used
            if lfirst_cmp then
                if loutradwind or loutdbz then
                    Call calc_geometry
                end if
                lfirst_cmp = .false.
            end if
            if loutradwind or loutdbz then
                if loutradwind then
                    Call calc_mod_radialwind(time)
                end if
                if loutdbz then
                    Call calc_mod_reflectivity(time)
                end if
                Call output_radar
            end if
        end if
    end if
end for
```

5. Sensitivity experiments

In order to specify appropriate configurations for the purpose of operational data assimilation, a series of sensitivity experiments are conducted in a module-wise way, i.e., we begin with the experiment E_0 associated with the simplest (or probably fastest) formulations for each module and upgrade one particular module in each successive experiment (see Tab. 5.1). We attempt to find an optimal configuration in the sense of balance between physical accuracy and computational expense by comparing the results of experiments with observations, because it provides us the opportunity to verify which formulation leads to significant errors and which does not, or in the other words, whether a more comprehensive formulation is necessary. With a clear physical understanding, we would like to do the least possible amount of adjustment in the operator to improve the comparisons.

All experiments are run on 16 processors (on a single node) of the NEC SX-9 cluster at DWD. The horizontal resolution of the COSMO-model is set to be $(0.025°, 0.025°)$ and vertical grid spacing ranges from 20 m at the bottom to 1000 m on the top in 51 levels. The one-moment microphysics schemes are used and the initial and boundary conditions are provided by the COSMO-EU model. For comparison, observations come from the radar network of DWD. Recall that these radars have 18 elevations, 1° azimuthal resolution, 1 km range resolution and 124 km effective range. To avoid the overlapping data issue caused by taking measurements from multiple radars, a single radar is involved in each case study.

As for quantitative comparisons between observations by radar and model simulations, the Contoured Frequency with Altitude Diagrams (CFADs, Yuter and Houze (1995)) are used, which summarize the vertical structure of the radar echoes through the frequency distribution of three-dimensional gridded reflectivity data and provide insight into the microphysical processes and structure of precipitating cloud systems. Yuter and Houze (1995) analyzed the transition of convective to stratiform precipitation in CFADs and pointed out that radar volume scans taken in stratiform event

Experiment	Pulse volume averaging	Beam bending	Scattering Schemes	Attenuation	Reflectivity weighting
E_0	No	43ERM	Rayleigh	No	No
E_1	No	SODE	Rayleigh	No	No
E_2	Yes	SODE	Rayleigh	No	Yes
E_3	Yes	SODE	Mie	No	Yes
E_4	Yes	SODE	Mie	Yes	Yes

Tab. 5.1.: Configuraions of the sensitivity experiments

showed narrow reflectivity distributions at all altitudes, while convection accompanied with broader distributions, especially in early stages. Therefore, in view of different meteorological situations, a stratiform precipitation case on 19 January 2012 and a convection case on 30 June 2012 are explored below.

5.1. The 19 January 2012 stratiform precipitation event

Stratiform precipitation is usually caused by large-scale, dynamic ascent of stably stratified, saturated air and is characterized by relatively small vertical air motion and large horizontal homogeneity of precipitation fields. It may extend for hundreds of kilometer but produce in general low rain rates that rarely exceed 10 mm/h. Precipitation typically first forms at high levels in the atmosphere. In this region, the temperature is normally below the freezing point of water. The precipitation forms initially as a succession of very small ice particles. Owing to weak upward air motions, ice particles of all types drift downward. During the sedimentation phase, they collect each other to form large snowflakes. As the snowflakes fall, they muss pass through the melting (or 0°) level where the temperature rises above the freezing point. At this stage the snowflakes will start to melt. The initial melting will be on the exterior of the snowflake where a water coating develops. Water is about 9 times more reflective than ice at microwave wavelengths, so these large wet snowflakes exhibit a high reflectivity. The highly reflective melting snow appear to the weather radars as more intense than it actually is. As the melting snow continues to fall and melt until it finally becomes rain drops that are smaller and fall faster, so both the size of the drops and their concentration are reduced, reducing the reflectivity. All of these processes lead to the formation of a narrow area of high reflectivity near the melting level, called the "bright band", which is considered

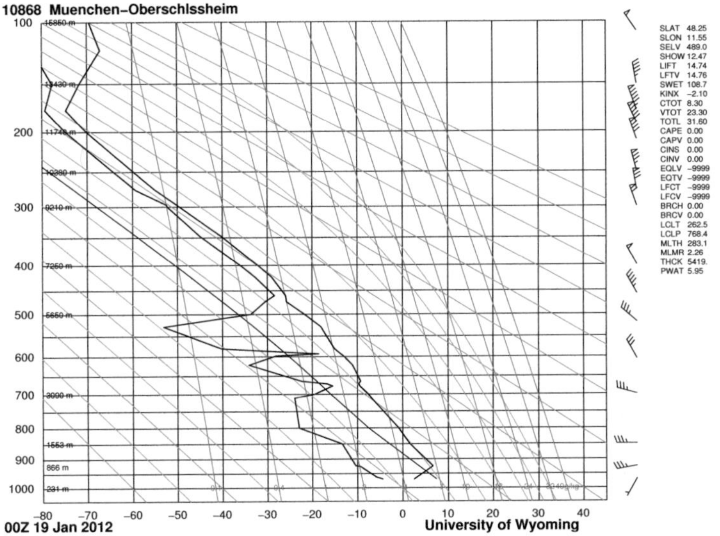

Fig. 5.1.: Sounding from Munich-Oberschleissheim at 00:00 UTC on 19 January, 2012

a major source of error in precipitation estimation. However, in a stratiform system the formation of heavily rimed ice species like graupel/hail is quite unlikely, so that less complicated microphysical processes are involved, which provides ideal conditions for the preliminary evaluation of microphysical parametrization schemes in snow and rain as well as the representation of the bright band in the operator/model. Commonly, the model is expected to agree with a stratiform precipitation event to a high extent.

5.1.1. Description of weather conditions, model data and observations

On 19 January, two low pressure systems "Fabienne I and II" moved from North sea to Baltic. To the south of the pressure centers, large-scale warm, cold and occluded fronts took place alternately in Central Europe, which caused widespread heavy rainfall. In the early morning, west wind brought warm Atlantic air into southern Germany and lifted rapidly the snow line up to about 1.8 km (see Fig. 5.1). Therefore, except the highest altitudes of Black Forest and Bavarian Forest where snow or sleet could still occur, heavy rain fell in most part of southern Germany, which caused several river floods in the northern part of Bavaria.

With respect to the study area, the radar station of Munich-Fuerholzen is chosen. According to Tab. 5.1, five COSMO-model runs are done, starting at 00:00 UTC on 19 January 2012 when the stratiform system entered the study area until 06:00 UTC. The initial and boundary conditions for model runs are interpolated from the COSMO-EU model at 00:00 UTC on 19 January 2012. The observed and simulated reflectivities are written in two output files, respectively, with a temporal resolution of 15 minutes.

5.1.2. Observed and simulated evolution of the precipitation

Firstly, we focus on examining the model representation of the precipitation in aspects of the timing, intensity and spatial distribution of the reflectivity by means of comparison of observed and simulated 2.5° PPI radar scans in 3-hour intervals. Later in this section, we will also specify and analyze the differences among simulations.

Fig. 5.2 views the observed and simulated 2.5° PPI scans at 00:00 UTC. In the observations, the event just moves into the study area from the northwest and the highest reflectivities reach 25 dBZ. Those features are fairly well captured in the simulations.

At 03:00 UTC (Fig. 5.3), the precipitation has arrived over the radar site and has been spread out over a larger area and the maximum values of reflectivities (appr. 40 dBZ) is found beneath the melting level, so it can be explained by the brightband effect. In the simulations, the position and intensity of the event are well represented and the maximum values of simulated reflectivities can also be seen below the melting level, however, approximately 5 dBZ less than observed.

At 6:00 UTC (Fig. 5.4), the precipitation persists over the radar site and the intensity has grown further with more reflectivities exceeding 40 dBZ, mainly distributed in the melting layer, below which reflectivity tends to decreases with the height. In contrast, the brightband structure is less clear in the simulations, the highest reflectivities attain just 35 dBZ and simulated reflectivity remains roughly constant below the melting level.

It is also noteworthy to mention that an evident overestimation of simulated spatial distribution in the higher atmosphere due to the presence of a huge amount of simulated reflectivities with values $\leq -5\,\text{dBZ}$ can be seen throughout the whole case study period. This is because the threshold value for the simulated reflectivity (-90 dBZ) is set much lower than that for radar measurements (about -31.5 dBZ).

5.1.3. CFADS and effects of operator modules

To help further understand some of the behaviors seen in the previous figures, CFADs are constructed for observed and simulated reflectivity (see Fig. 5.5). The different colors represent the percentage of particular reflectivity values falling in a given class compared to the total number of reflectivity values above a given threshold ($\geq -30\,\text{dBZ}$) at a given height. The class size selected for this diagram is 2 dBZ. The contours are at interval of 2% per dBZ per 500 m.

In Fig. 5.5, CFADs for the observations and five simulations at 06:00 UTC are shown. In a gross sense, Fig. 5.5a characterizes the typical stratiform precipitation as observed: the highest probabilities follow a coherent pattern with the peak density continually decreasing with the height from about 25 dBZ around the melting layer (roughly at 2 km high) to -10 dBZ around 7 km and remaining constant afterwards. Below the melting layer, peak probabilities decrease slightly, probably due to the evaporation, and then remain constant down to the surface. Maximum reflectivities are slightly over 35 dBZ at the surface, close to 40 dBZ around 2 km, and decay steadily upwards until about 7 km.

Although both in observation and simulations the frequency distributions of reflectivity at upper levels are generally different from those at low levels, notable discrepancies can be seen while all five simulations are quite similar to each other. In CFADs of the simulations, reflectivity has a more rapid linear decrease above the melting level, which leads to more than 10 dB underestimation of reflectivity on average around 7 km compared with observations. The distributions below the melting level are much narrower and show no evidence of decrease by evaporation. Near the surface, the maximum simulated reflectivities are slightly lower than observed, which may be explained by the remaining ground echos not removed by the clutter removal algorithm for the observed reflectivity. However, 30 dBZ echoes occur at a frequency of more than 20% compared to about 2% in observation. Since the raindrop collision/coalescence process plays a relatively small role in shaping its size distribution due to the low rainfall rates

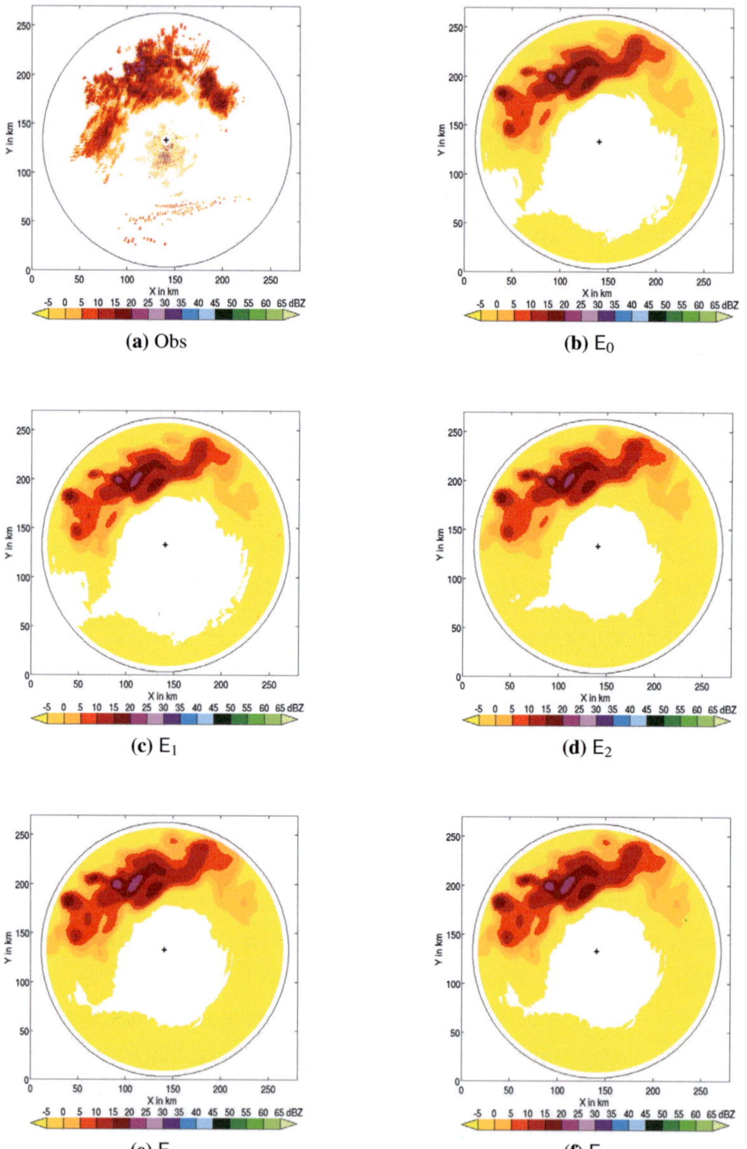

(a) Obs (b) E_0 (c) E_1 (d) E_2 (e) E_3 (f) E_4

Fig. 5.2.: Radar reflectivity in dBZ (see color bar) at an elevation of 2.5° (PPI mode) on 19 Jan, 2012, 00:00 UTC: (a) observations; (b)-(f) different sensitivity results (see details in text)

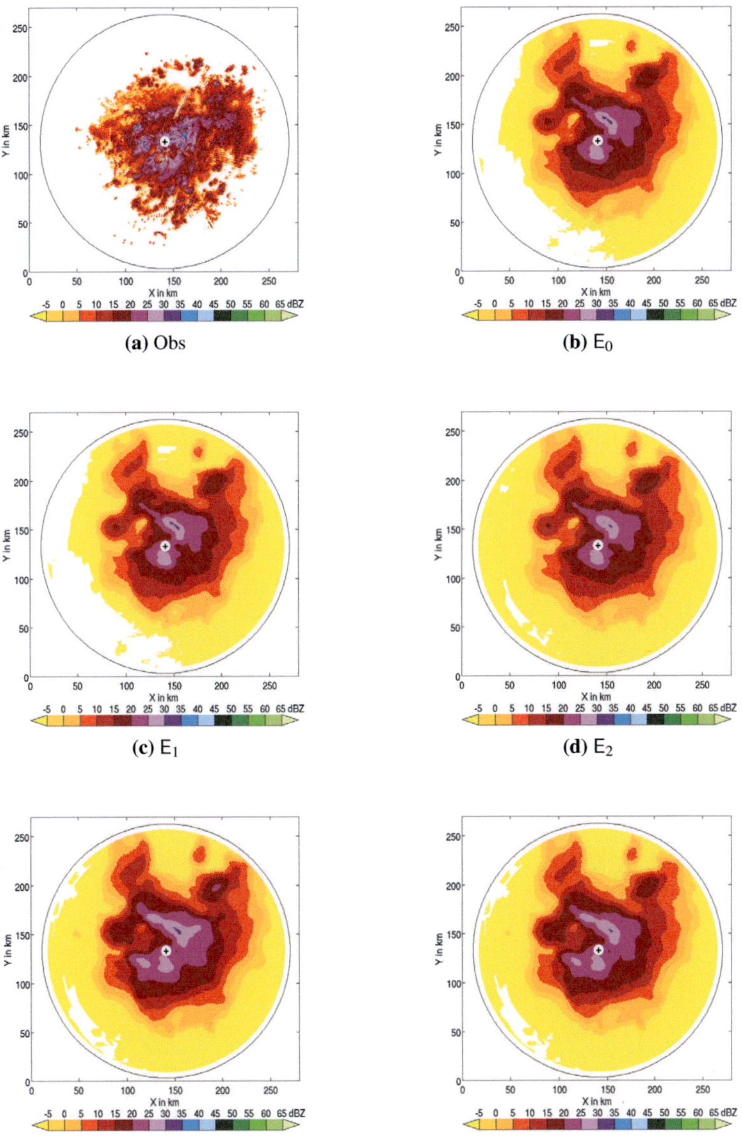

(a) Obs

(b) E_0

(c) E_1

(d) E_2

(e) E_3

(f) E_4

Fig. 5.3.: As Fig. 5.2 but for 03:00 UTC

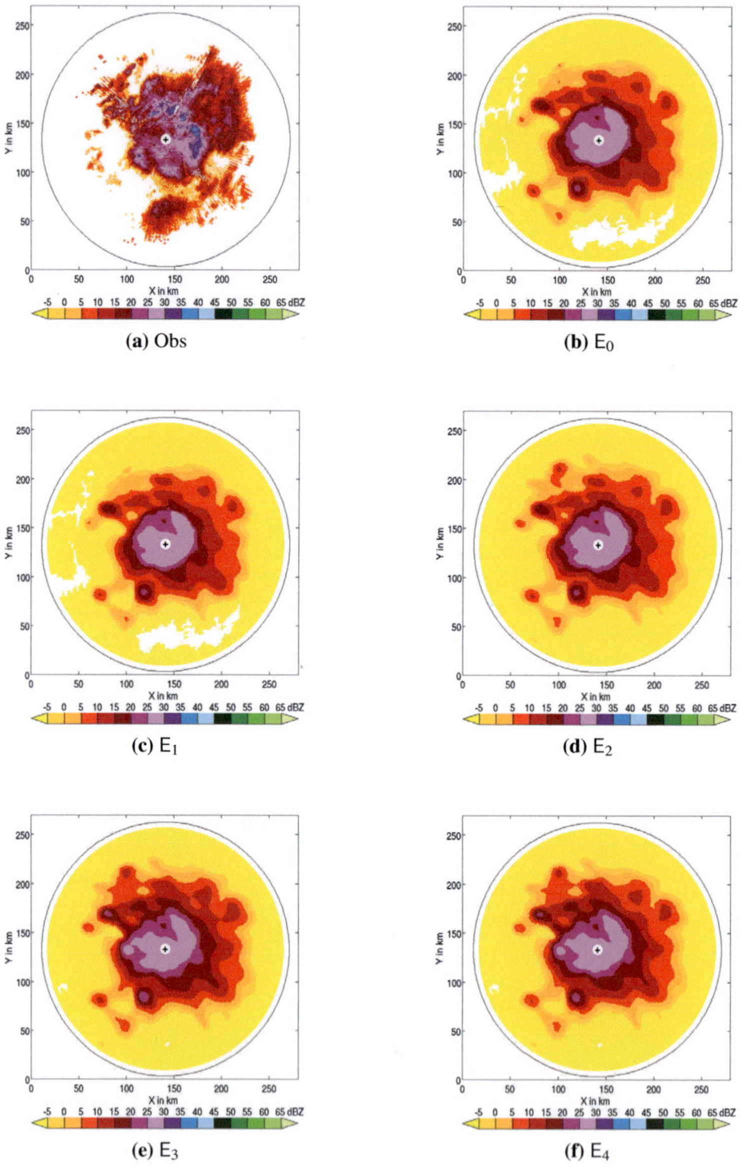

Fig. 5.4.: As Fig. 5.2 but for 06:00 UTC

(generally less than 6 mm/h) in stratiform precipitation, the high concentrations of simulated large raindrops near the surface mainly arise from melting large ice-phase particles, as indicated by high frequency of reflectivities ranging from 10 to 20 dBZ between 2 and 3.5 km.

In addition to examining the behavior of simulations from the CFAD perspective, this study also attempts to verify the performance of the individual configurations of the operator. Here we take for instance the 2.5° PPI at 06:00 UTC on 19 January 2012 (Fig. 5.7), which shows some important differences. The effects of upgrades are depicted in differences between two experiments, denoted as $\mathsf{E}_{i+1} - \mathsf{E}_i$ (the subtraction is class-wise). The classes, at which either the observed or the simulated data are not present and thus the amount of differences can not be expressed, will be marked in color black. As shown in Fig. 5.7a, visible differences occur at a distance of about 100 km where height differences computed by 43ERM and SODE become considerable, say 100 m (Fig. 5.6b). In Fig. 5.7b, the clearest differences (in dark red) appear at further distances, resulting from comparisons of negative reflectivities, and the black area indicates a larger areal coverage by means of pulse-volume averaging. However, differences in radius of 30 km, that is, about 2 km in height, are more interesting. By virtue of the height, the differences arise probably from the beam interception of the melting level and consequently pulse-volume averaging over an inhomogeneous vertical profile of reflectivity around the melting level. In Fig. 5.7c, it can be seen that the Mie-scattering scheme produces higher reflectivities than the Rayleigh's method regarding the melting particles between heights of 30 and 70 km (cf. Section 3.7) but no notable differences when it comes to light snow and large rain drops in the lower and upper atmosphere. Finally, Fig. 5.7d shows that attenuation does not play an essential role in stratiform precipitation.

Experiment	Operator [s]	Total model run [s]	Ratio [%]
E_0	72.10	348.24	20.70
E_1	81.70	354.65	23.04
E_2	117.36	387.70	30.27
E_3	143.79	419.71	34.26
E_4	150.59	426.59	35.30

Tab. 5.2.: Elapsed wall-clock time: (Ratio = Operator/Total model run $\times$ 100)%

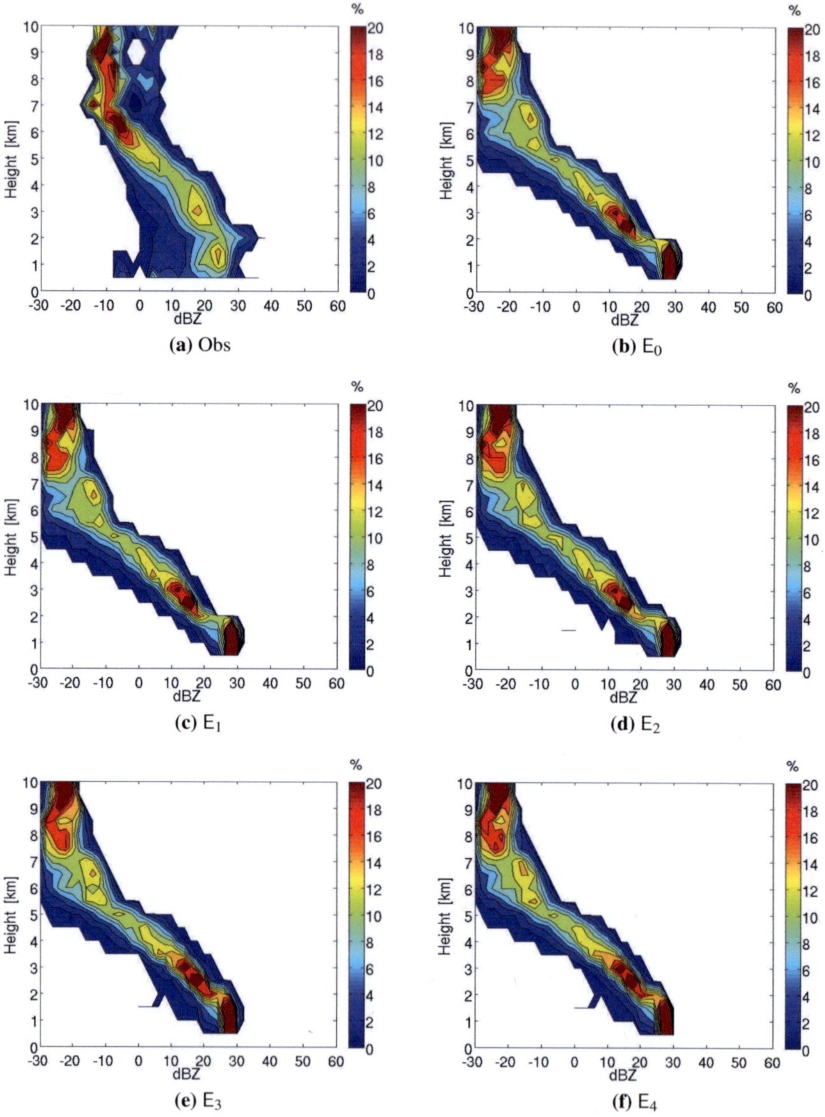

Fig. 5.5.: CFADs of radar reflectivity in % on 19 January, 2012, 06:00 UTC: (a) Observed; (b)-(f) different sensitivity results (for details see text)

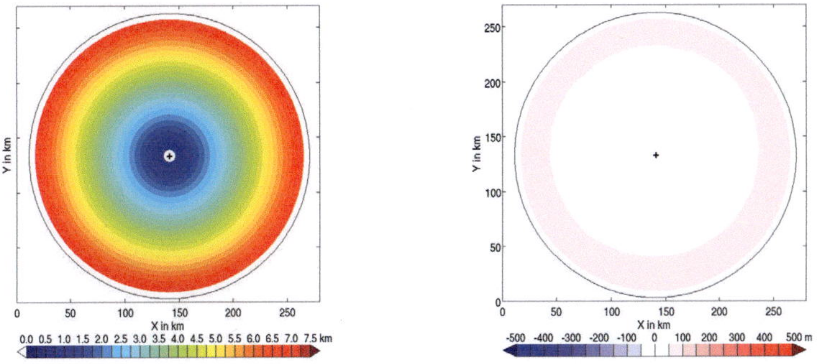

(a) E_0 **(b)** $E_1 - E_0$

Fig. 5.6.: Left column: height in km (see color bar) calculated by 43ERM (from E_0), at an elevation of 2.5° (PPI mode) on 19 January, 2012, 06:00 UTC; **Right column:** height difference in m (see color bar) between SODE (from E_1) and 43ERM

As aforementioned, efficiency of the operator is also regarded as a crucial criterium for performance. Tab. 5.2 lists the elapsed wall-clock times of total model run and operator for each experiment. We should be aware that there are variable number of background computations on SX-9 from time to time, which may delay the model runs. Typically, operational routines of DWD with higher priority can drive away our runs on the waiting queue. So the absolute time differences depicted in Tab. 5.2 do not mean definitely how much one run is faster/slower than the others, however, it does show us that the time expense tends to ascend with updating configurations, but not dramatically. The operator occupies minimal 20.70% time of total model run in E_0 and maximal 35.30% in E_4. The most significant increase occurs when pulse-volume averaging is switched on in E_2, followed by E_3 that puts Mie-scattering scheme in action.

Tabs. 5.3-5.7 reveal the specific elapsed wall-clock time distribution of the operator in each experiment. We consider the sum of time consumed by "Init./const. geom." (Algs. 1 and 6), "Grid point values" (calculation reflectivity, attenuation coefficient and fall speed on model grid), "Online beam propag." (Alg. 13) and "Comp. on polar grid" (Algs. 3, 4, 7, 9, 10, 15 and 16) as work load and the others, i.e., "MPI Communications" (Alg. 12) and "Output" (Alg. 5) as communication overhead. The

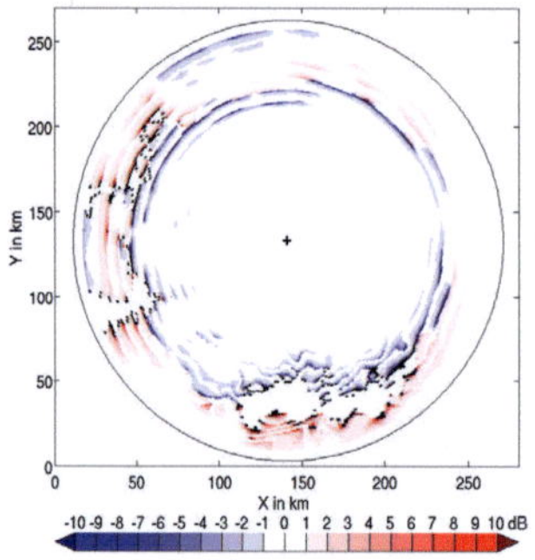

(a) $E_1 - E_0$

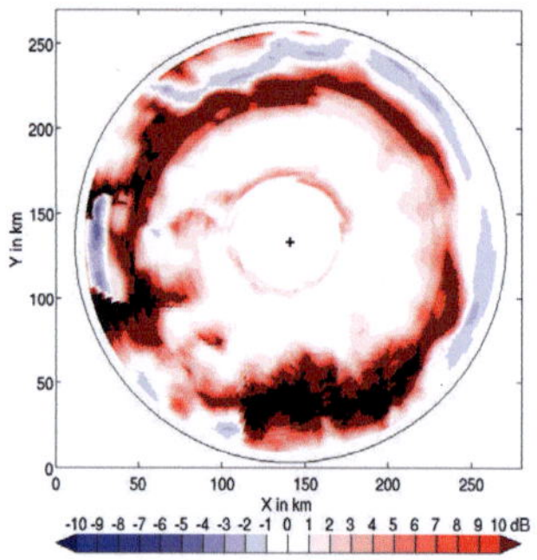

(b) $E_2 - E_1$

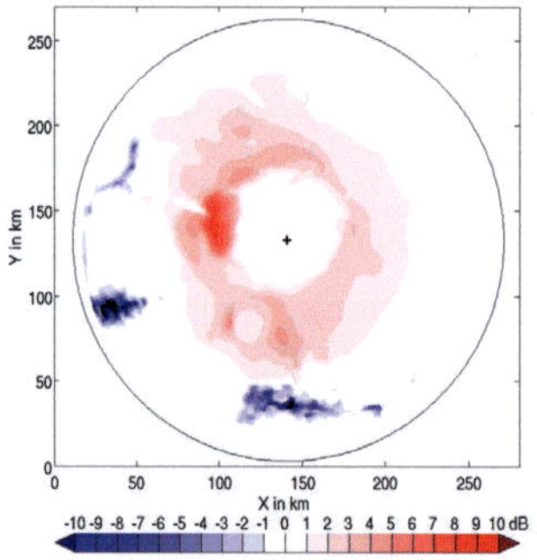

(c) $E_3 - E_2$

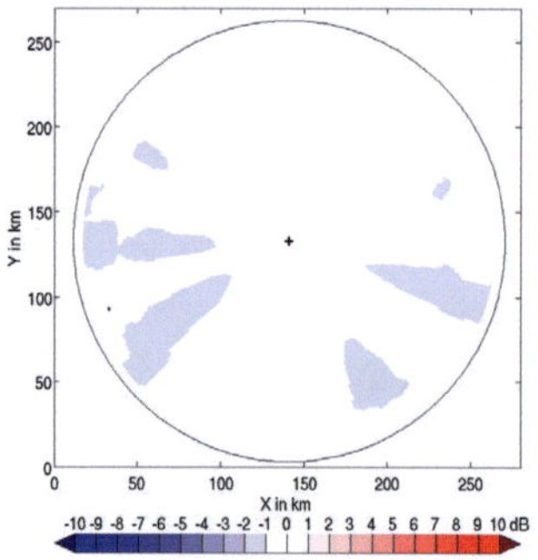

(d) $E_4 - E_3$

Fig. 5.7.: Reflectivity difference in dB (see color bar) of sensitivity results at an elevation of 2.5° (PPI mode) on 19 January, 2012, 06:00 UTC modest time growth in E_1 arises from the online simulation of beam propagation and the associated communication efforts. Due to intensive pulse-volume averaging, that is $(nh, nv) = (5,9)$, E_2 has obviously much more ray paths to simulate in "Online beam propag." and additional integration steps in "Output". The main increase in E_3 stems from "Grid point values", where considerable time has to be spent in establishing a lookup-table for the Mie-scattering scheme. Fortunately, this work needs to be done just once, so it will become less immaterial for model runs of long term. The slight increase in E_4 results from the sum operations of attenuation coefficients to attain total attenuation in "Output".

With respect to the topic of load balance, which is defined here as the average work load among all processors divided by the maximum work load (the ideal case is a load balance of one and the worst case is the reciprocal of the number of processes), the operator exposes a proficient load balance, indeed, above 97% (see Tab. 5.8). The "lowest" load balance in E_0 can be explained by the inhomogeneous distribution of clouds in the model domain, so that some processors have more to do than the others, and the load balance is continuously ameliorated with configuration updating.

Task	Min	Ave	Max
Operator	69.67	72.10	74.07
Init./const. geom.	12.08	12.18	12.29
Grid point values	1.98	4.23	6.13
Online beam prop.	0.00	0.00	0.00
Comp. on polar grid	0.00	0.03	0.25
Work load	14.06	16.44	18.67
MPI Comm.	0.04	0.70	0.95
Output	54.79	54.84	55.55
Comm. overhead	54.81	55.54	56.50

Tab. 5.3.: Elapsed time distribution in E_0

Task	Min	Ave	Max
Operator	79.16	81.70	83.76
Init./const. geom.	12.66	12.78	12.90
Grid point values	1.98	4.38	6.34
Online beam prop.	0.57	0.58	0.59
Comp. on polar grid	0.06	0.09	0.17
Work load	15.27	17.83	20.00
MPI Comm.	5.64	6.33	6.60
Output	57.83	57.43	58.14
Comm. overhead	63.47	63.76	64.74

Tab. 5.4.: Elapsed time distribution in E_1

Task	Min	Ave	Max
Operator	114.94	117.36	119.40
Init./const. geom.	12.38	12.47	12.59
Grid point values	2.05	4.33	6.26
Online beam prop.	14.67	15.02	15.53
Comp. on polar grid	1.13	1.19	1.31
Work load	30.23	33.01	35.69
MPI Comm.	6.00	7.20	7.68
Output	76.95	77.03	77.72
Comm. overhead	82.95	84.23	85.40

Tab. 5.5.: Elapsed time distribution in E_2

Task	Min	Ave	Max
Operator	141.31	143.79	145.84
Init./const. geom.	12.22	12.33	12.44
Grid point values	24.06	26.17	28.04
Online beam prop.	14.77	15.21	15.60
Comp. on polar grid	1.13	1.19	1.30
Work load	52.18	54.90	57.38
MPI Comm.	6.30	7.38	8.03
Output	81.33	81.40	82.09
Comm. overhead	87.63	88.78	90.12

Tab. 5.6.: Elapsed time distribution in E_3

Task	Min	Ave	Max
Operator	148.14	150.59	152.57
Init./const. geom.	12.26	12.36	12.48
Grid point values	23.89	26.05	27.99
Online beam prop.	14.62	15.00	15.55
Comp. on polar grid	1.42	1.50	1.63
Work load	52.19	54.91	57.65
MPI Comm.	7.38	8.90	9.63
Output	86.58	86.66	87.49
Comm. overhead	93.96	95.56	97.12

Tab. 5.7.: Elapsed time distribution in E_4

Experiment	Load balance [%]
E_0	72.10/74.07 = 97.34
E_1	81.70/83.76 =97.54
E_2	117.36/119.40 = 98.29
E_3	143.79/145.84 = 98.59
E_4	150.59/152.57 = 98.70

Tab. 5.8.: Load balance

5.2. The 30 June - 01 July 2012 convective precipitation event

The dynamical characteristics of convective processes differ stronly from those of stratiform events, which results in distinction in microphysics and spatial distribution of precipitation. In meteorology, convection refers primarily to heat transport by vertical motions of the flow, being produced by differences in bouyancy arising from variations in density. Vertical motions are about 1 to 10 m/s or more, which equals or exceeds terminal fall velocities of ice particles (Houze, 1993), so that particles are rapidly carried up and down inside the cloud by up- and downdrafts and grow by riming, which allows the formation of larger ice species like graupel or hail. Because updrafts exist in a limited region of the convective clouds, radar echoes associated with active convection form a vertical region of maximum reflectivity, which contrasts with the horizontal orientation of radar bright band seen within the melting layer of stratiform precipitation. The current COSMO-model is supposed to be able to resolve convective systems, but to which extent a convection can be outlined by the event and the model will be investigated in what follows.

5.2.1. Description of weather conditions, model data and observations

Massive warm and humid air had accumulated in the south of Germany for several days before 30 June 2012. At midnight, the temperature was exceptionally high in Bavaria (see Fig. 5.8). Together with the high humidity, it provided the necessary energy for the severe nighttime storm. As a low pressure system expanded reluctantly from Southwestern Europe to Bavaria on the night of 30 June, strong thunderstorms were ultimately triggered, accompanied with heavy rain, hail and stormy southwest winds. The hail was mostly small-grained, but some of which reached diameter up to 4 cm.

In this case, the simulations are initialized using the COSMO-EU model at 21:00 UTC, 30 Juni 2012 and run for 4 hours. In addition to reflectivity, Doppler velocity are simulated and analysed as well. Observations are collected from the radar site Türkheim, located at the Swabian Alb.

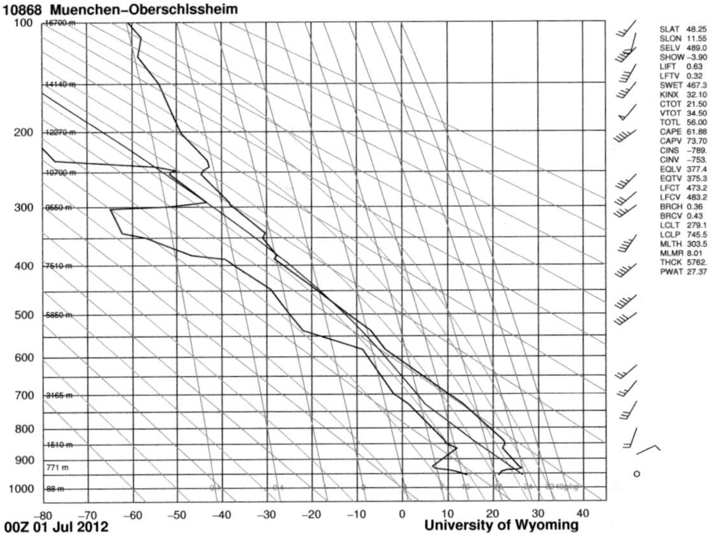

Fig. 5.8.: Sounding from Munich-Oberschleissheim at 00:00 UTC on 01 July, which is about 90 km away from Türkheim.

5.2.2. Observed and simulated evolution of the precipitation

In analogy to the stratiform case study, the evaluation of the representation will be based on 2.5° PPI radar scans compared to five simulations each two hours starting 21:00 UTC on 30 Juni 2012 until 01:00 UTC on 01 July 2012 when the convective system had almost totally faded in the study area. Because of the complex microphysical processes in convection, discrepancies in timing, organization and intensity of the event are expected.

Fig. 5.9 and 5.10 show the observed and simulated 2.5° PPI scans at 21:00 UTC on 30 June 2012 for reflectivity and Doppler velocity, respectively. In the observations (Fig. 5.9a), the convective system propagates to the study area from the northwest with the main squall line in the west (with N/S orientation) spreading from 150 km to 250 km in Y direction. The line has the highest reflectivities up to 55 dBZ in the south and weakens gradually towards the north. From the north to the northeast, the observations are dominated by reflectivities between 20 and 30 dBZ with some convective cells in the vicinity of (230, 150) km. All the simulations (Figs. 5.9b-5.9f) are able to rebuild the movement and shape of the event appropriately but the intensity is comparable apart from observations. In the northern part, the number of reflectivities between 20 and 30 dBZ are mainly overestimated in simulations and the position of observed convective cells around (230, 150) km seems to be shifted to the northwestern sector near the radar site. The simulated reflectivities of value ≤ -5 dBZ are mostly distributed in the higher atmosphere, while the observed ones are concentrated in the lower levels. In terms of Doppler velocity, all the simulations show a good agreement with the observations in strength and direction. Since Doppler radars can only measure wind where reflectivity is present, the observed wind shows an identical areal extent as reflectivity. Ignoring this factor, E_0 and E_1 overestimate clearly the coverage of Doppler wind compared to the observations, especially in the southern part, where no reflectivity is observed (see Fig. 5.9a). Fig. 5.10d highlights the effects of taking weighting by reflectivity into account, which adjusts the coverage close to the observed one. In Figs. 5.10e and 5.10f, no distinction can be seen, though.

At 23:00 UTC, the squall line has arrived in Türkheim, the highest reflectivities of which have strengthened slightly with maximal value up to 60 dBZ, as shown in Fig. 5.11a. In the simulations, the system has also moved further to the east but

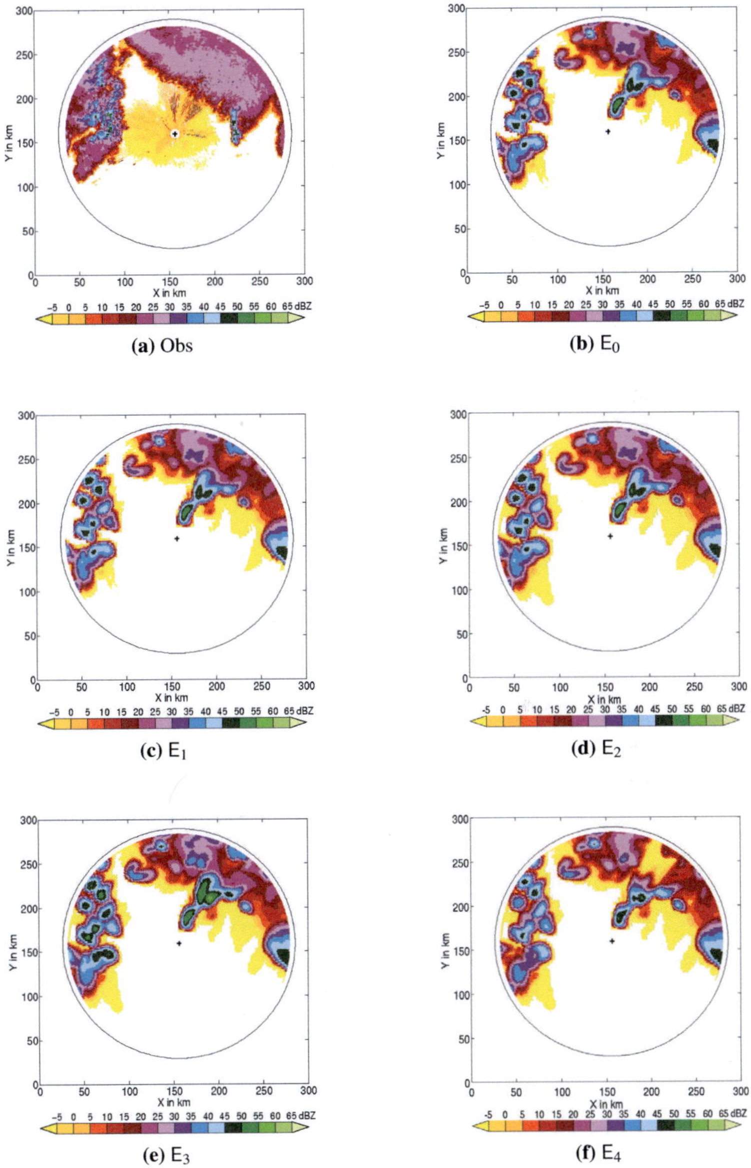

(a) Obs (b) E_0 (c) E_1 (d) E_2 (e) E_3 (f) E_4

Fig. 5.9.: Radar reflectivity in dBZ (see color bar) at an elevation of 2.5° (PPI mode) on 30 June, 2012, 21:00 UTC: (a) observations; (b)-(f) different sensitivity results (see details in text)

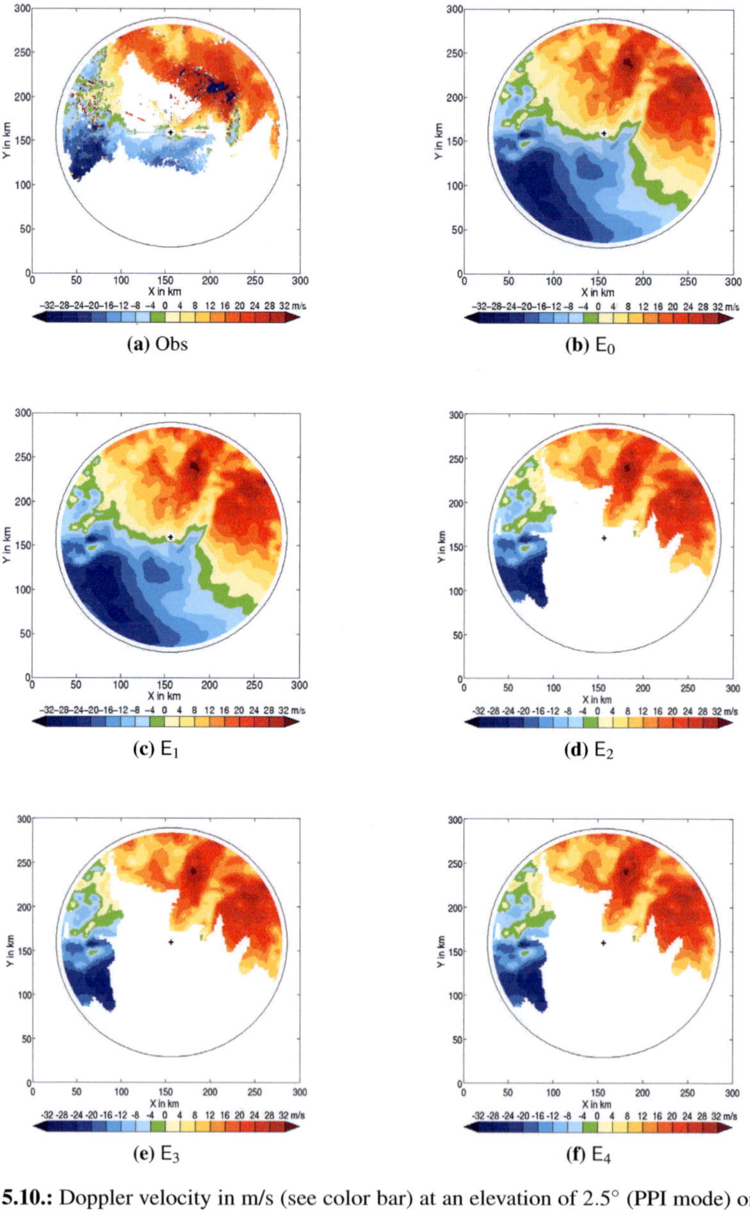

(a) Obs (b) E_0 (c) E_1 (d) E_2 (e) E_3 (f) E_4

Fig. 5.10.: Doppler velocity in m/s (see color bar) at an elevation of 2.5° (PPI mode) on 30 June, 2012, 21:00 UTC: (a) Observed; (b)-(f) different sensitivity results (for details see text)

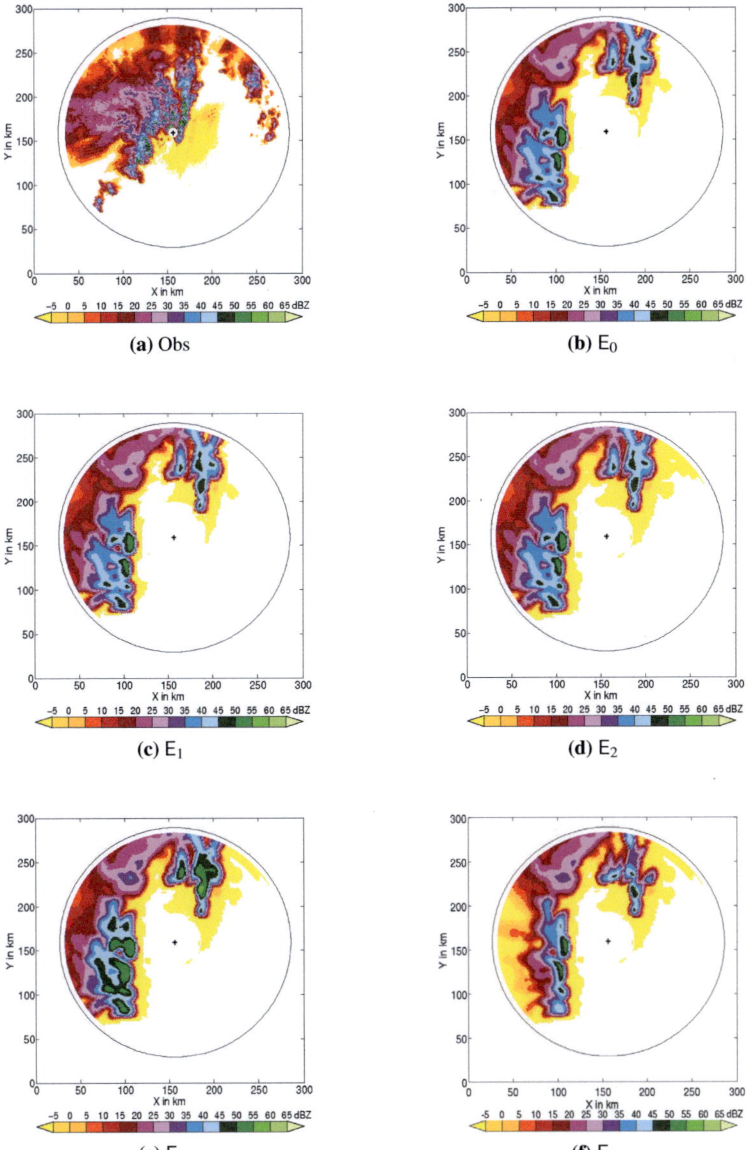

(a) Obs **(b)** E_0 **(c)** E_1 **(d)** E_2 **(e)** E_3 **(f)** E_4

Fig. 5.11.: As Fig. 5.9 but for 23:00 UTC

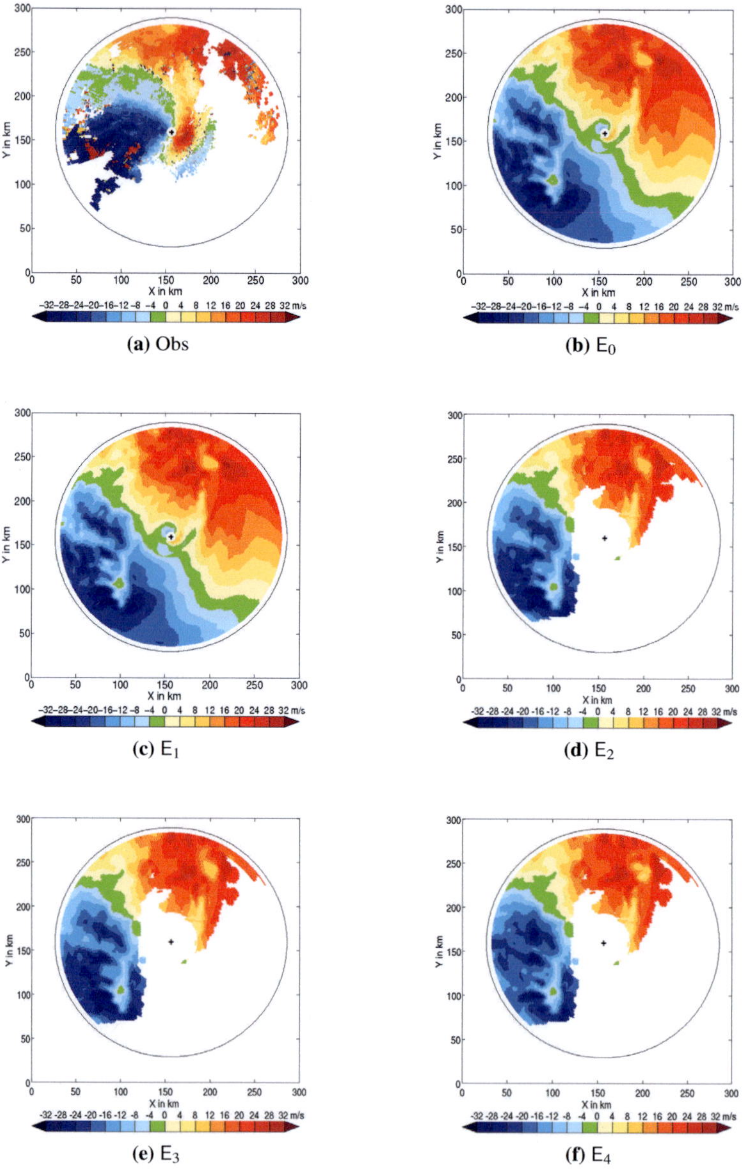

Fig. 5.12.: As Fig. 5.10 but for 23:00 UTC

apparently more slowly and the highest reflectivities attain values just close to 55 dBZ. Related to the discrepancies in reflectivity, the simulated Doppler velocities are also apart from the observations in strength and distribution pattern, as shown in Figs. 5.12a and 5.12d.

At 01:00 UTC on 01 July 2012, the convection has passed Türkheim and the intensity of the event has significantly declined with just a few cells exceeding 45 dBZ and the squall line expends from the southwest to the northeast. The overall structure and evolution of the event are well established in the simulations but the number of reflectivity with value ≤ -5 dBZ is underestimated in the lower and overestimated in the higher levels beside the observations. This disagreement is also visible for Doppler velocity (Fig. 5.14d).

In order to compare the observed and the simulated evolution of the precipitation in a more quantitative way, we now express the temporal evolution of the system at 15-minute intervals as histograms based on different reflectivity classes given in Tab. 5.9.

Reflectivity class	Interval [dBZ]	Precipitation
1	$(-\infty, 5)$	no
2	$[5, 15)$	very light
3	$[15, 25)$	light
4	$[25, 35)$	moderate
5	$[35, 45)$	heavy
6	$[45, \infty)$	very heavy

Tab. 5.9.: Different reflectivity classes

Fig. 5.15 is derived from the 2.5° PPI scans. As shown in Fig. 5.15a, a great amount of reflectivities fall in Class 1 both in observations and simulations, and an evident overestimation of simulated reflectivities can be seen throughout the entire study period. For Class 2, 3 and 4, there are more observed reflectivities and the numbers peak at 23:30 UTC, closely one hour earlier than the simulations, which confirms the previous finding that the evolution of model is likely slower. Fig. 5.16 is the histogram of all reflectivities, and shows nearly the same characteristics as Fig. 5.15 does, which means that the 2.5° PPI scan is a suitable representative for the whole event. To

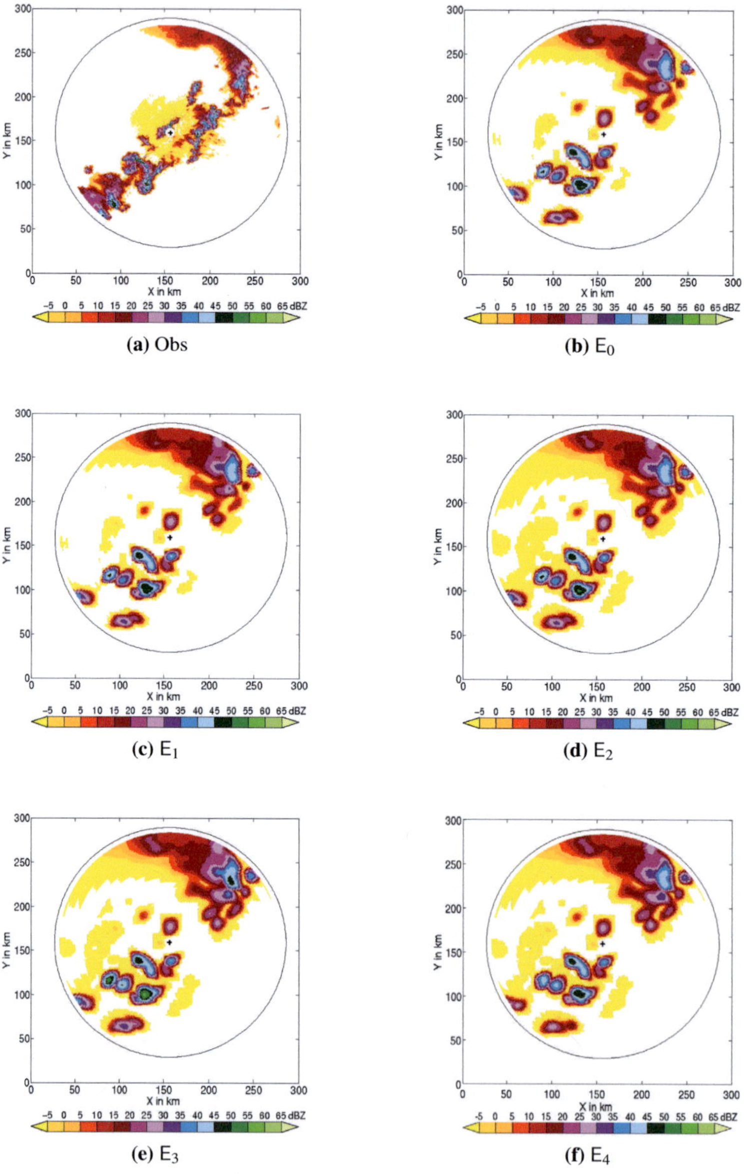

(a) Obs (b) E_0

(c) E_1 (d) E_2

(e) E_3 (f) E_4

Fig. 5.13.: As Fig. 5.9 but for 01 July, 2012, 01:00 UTC.

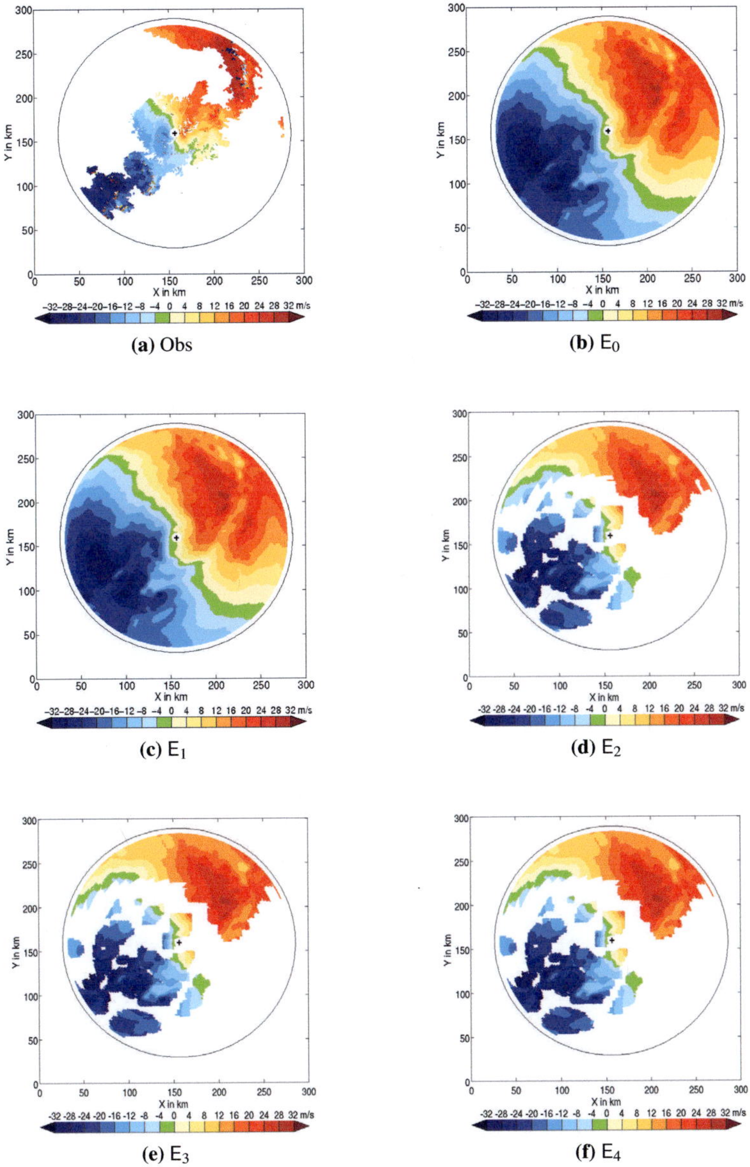

(a) Obs (b) E_0

(c) E_1 (d) E_2

(e) E_3 (f) E_4

Fig. 5.14.: As Fig. 5.10 but for 01 July, 2012, 01:00 UTC.

verify the model's delay, we add Fig. 5.17, the simulated PPI scan of elevation 2.5° at 00:00 UTC on 01 July 2012 for E_4, which shows that the reflectivity coverage and wind pattern in Figs. 5.17a and 5.17b are much closer to Figs. 5.11a and 5.12a than Figs. 5.11f and 5.12f. This lag is also visible in Class 5 and 6 although the number of simulated reflectivities is generally overestimated. The observations and simulations reach comparable numbers in all classed except Class 1 for the first time at 01:00 UTC on 01 July when the convective system has strongly decayed.

5.2.3. CFADS and effects of operator modules

Again for further analysis, reflectivity CFADs are constructed by classing the reflectivities into 2-dBZ classes beginning at -30 dBZ at each height. The contours here are at interval of 2% of data per dBZ per 1 km. Fig. 5.18 shows the observed and simulated CFADs for 23:00 UTC. In comparison to the stratiform case, the convective CFADs indicate much broader distributions of reflectivity values in all levels, which agrees with the statement of Yuter and Houze (1995). In the observed CFAD (Fig. 5.18a), above the melting level (appr. 3 km high), the intensity decreases with increasing height and high concentration of 20 dBZ reflectivities can be seen at 8 km height. Maximum reflectivities are around 55 dBZ at the surface, over 45 dBZ between 2 und 4 km height and drop off steadily aloft. High reflectivities ($\geq$ 40 dBZ) aloft suggest a predominance of graupel or hail, as consequence of convective updrafts favoring growth of particles by riming. The melting large ice-phase particles and the raindrop collision/coalescence process in convection give rise to the large raindrops under the melting level. In comparison, the shapes of observed and simulated distributions are considerably different, while all five simulations are similar to each other (Fig. 5.18). As indicated in Fig. 5.18b, the simulated reflectivities exhibit a even broader distribution and decrease faster with the increasing height than the observed ones down till the melting level, and the concentration of 20 dBZ reflectivities is not present. Below the melting level, most reflectivities accumulate at 40 dBZ, about 10 dBZ more than observations.

We give now a look at the effects of the individual configurations of the operator. Here we focus on the 2.5° PPI scans at 23:00 UTC. The differences are depicted in Figs. 5.19 (for reflectivity) and 5.20 (for Doppler velocity). As indicated in Fig. 5.19a, width=5.9 cm, height=6.6 cm

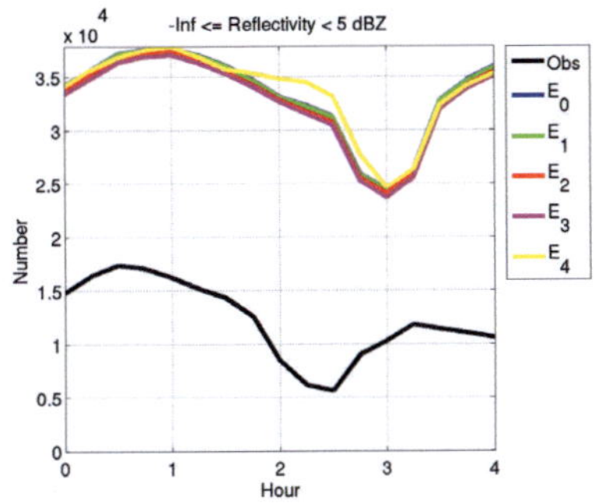

(a) Class 1

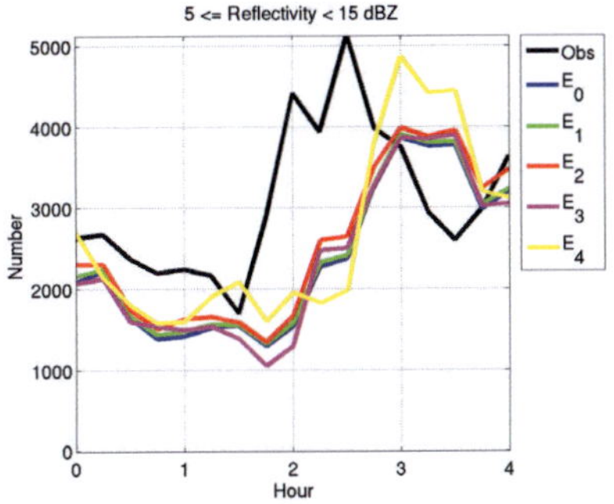

(b) Class 2

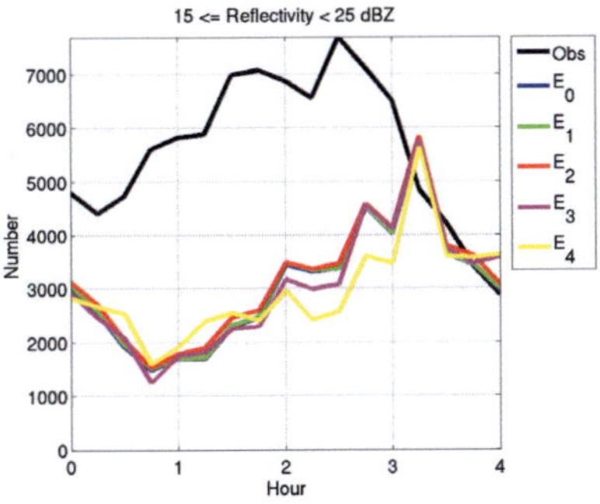

(c) Class 3

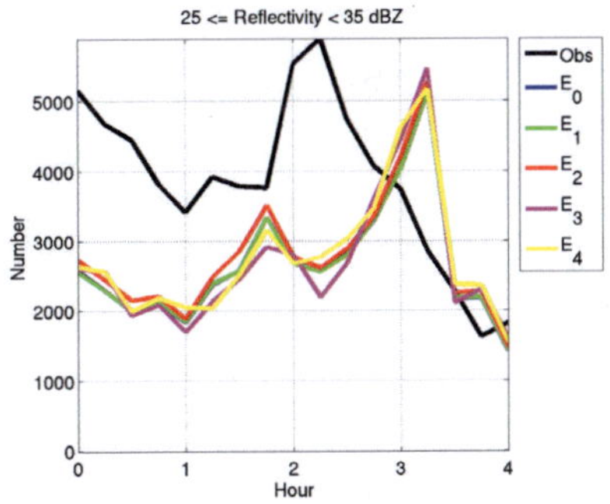

(d) Class 4

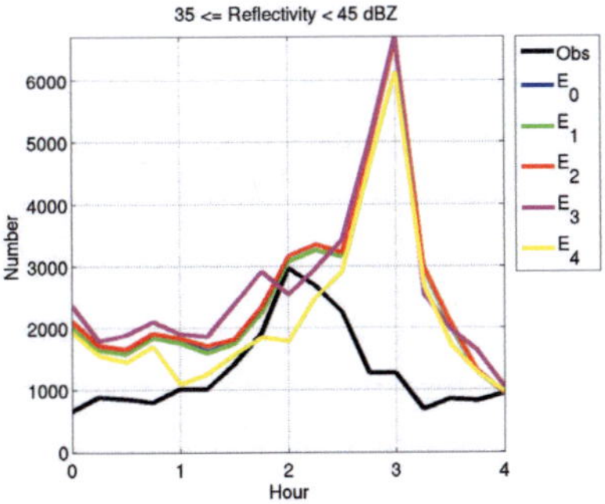

(e) Class 5

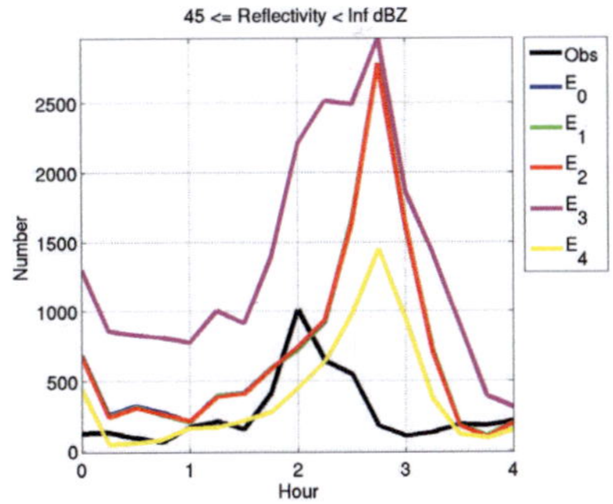

(f) Class 6

Fig. 5.15.: Histograms of reflectivity in different classes (a-f) as function of time at 15-minute intervals from 21 UTC on 30 June to 01 UTC on 01 July 2012 at an elevation of 2.5° (PPI mode): observations in black line; sensitivity results in color lines. The class intervals are written on the top of each subfigure. The X-axis is in units of hour, from 0 to 4 hours, representing from 21 UTC to 01 UTC. The Y-axis is the number of reflectivity bins.

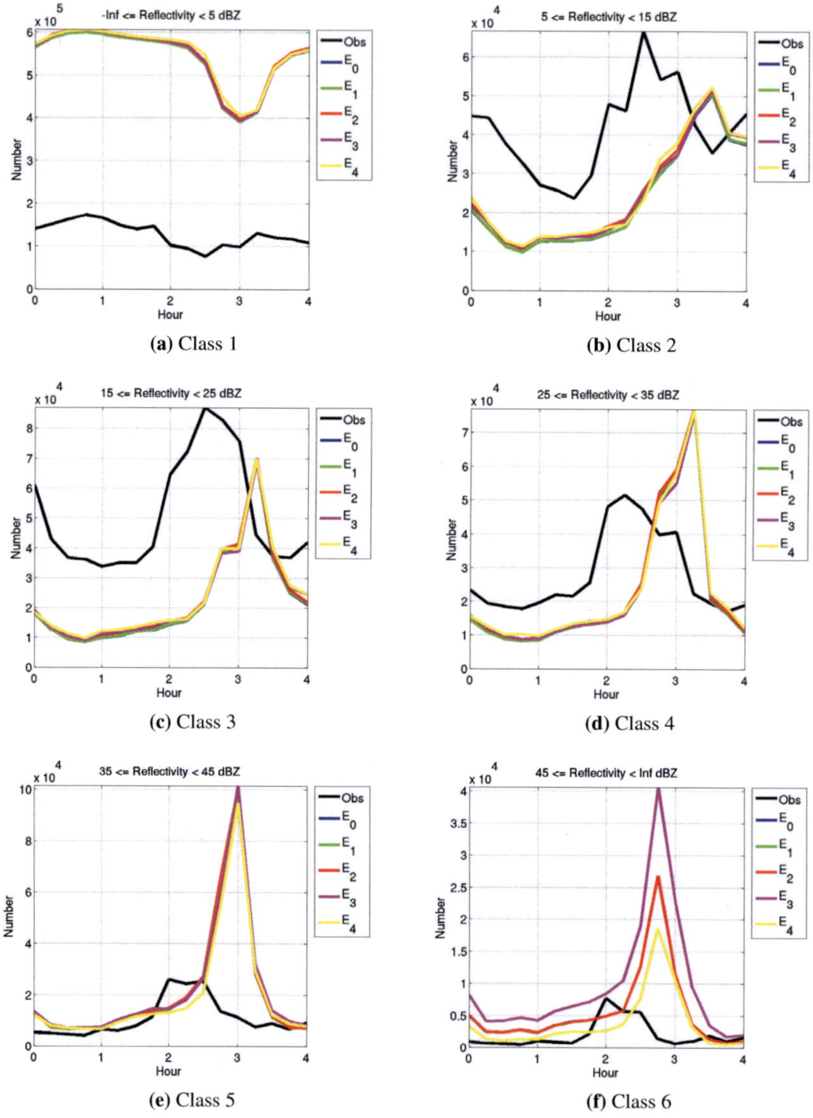

(a) Class 1

(b) Class 2

(c) Class 3

(d) Class 4

(e) Class 5

(f) Class 6

Fig. 5.16.: As Fig. 5.15 but for all elevations

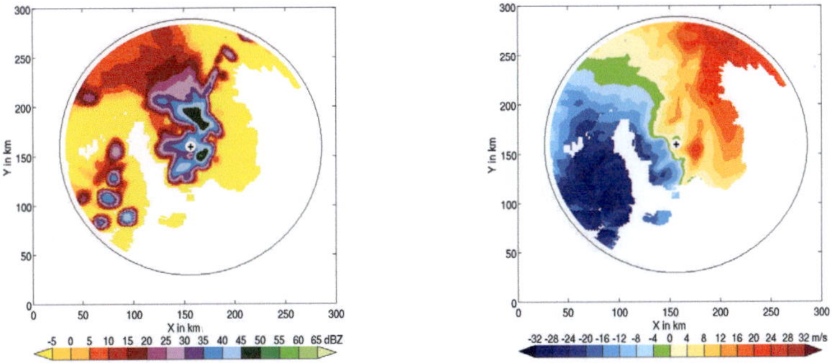

(a) Reflectivity **(b)** Doppler velocity

Fig. 5.17.: Sensitivity results of E_4: (a) radar reflectivity in dBZ (see color bar) at an elevation of 2.5° (PPI mode) on 01 July, 2012, 00:00 UTC; (b) Doppler velocity in m/s

the chosen methods for simulating beam propagation have trivial distinctions in this case. The differences in Figs. 5.19b and 5.19c arise from the reasons as in the stratiform case study. The most crucial impacts are probably when the attenuation is accounted for Fig. 5.19d, where the intensity of reflectivities weakens by more than 10 dB in the western and northern parts of the study area. For Doppler velocity, the most important improvement is brought about by weighting Doppler velocity by reflectivity (Fig. 5.20b). Besides the reduction of the spatial extent of wind to comparable surface of observations, the strength of wind also drops in general because reflectivity declines with height in most cases. The other upgrades reveal little importance.

5.3. Summary

To sum up this chapter, a good agreement between model and radar observations was achieved in the case of the stratiform event, including the intensity and position of the event and bright band. Regarding the convective event, the model is able to catch the general evolution of the system but with some time delay. This might have contributed to disagreements in intensity and position.

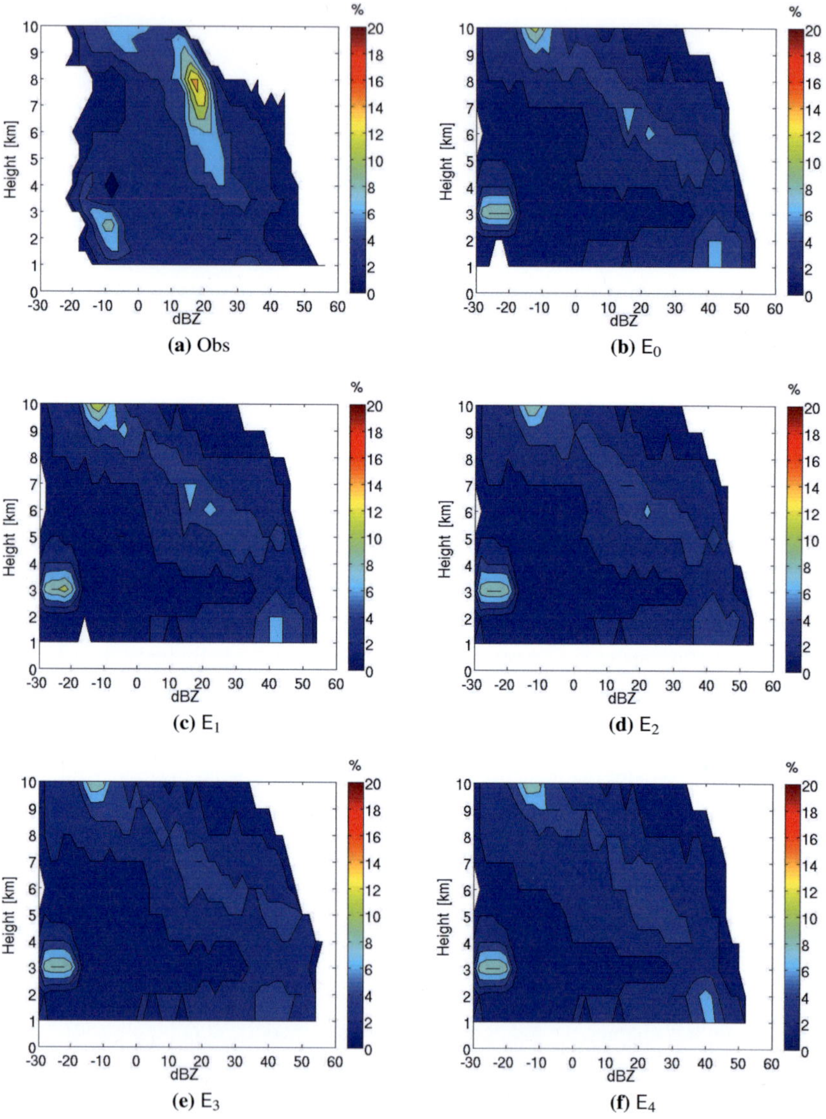

(a) Obs (b) E_0

(c) E_1 (d) E_2

(e) E_3 (f) E_4

Fig. 5.18.: As Fig. 5.5 but for 30 June, 2012, 23:00 UTC.

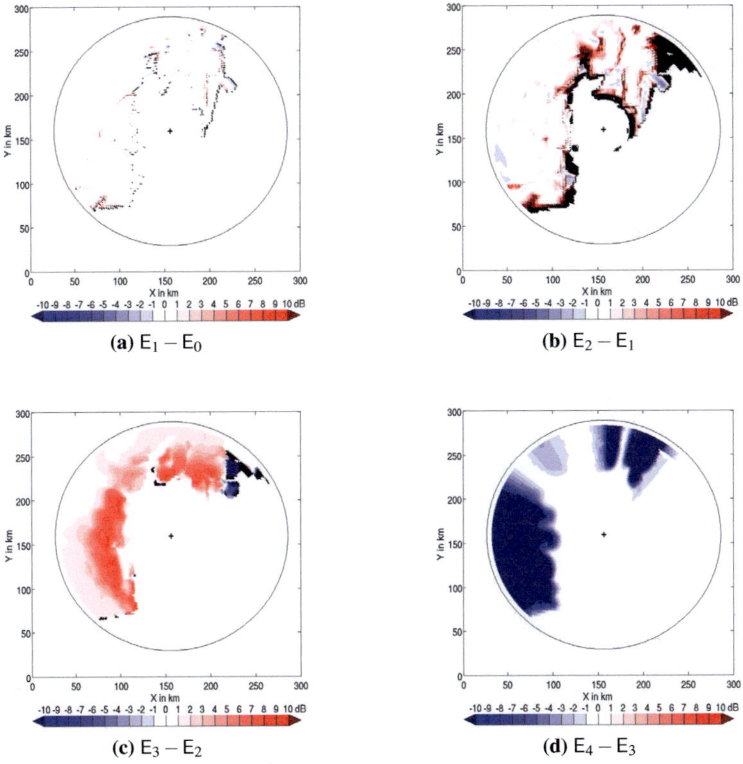

(a) $E_1 - E_0$ (b) $E_2 - E_1$ (c) $E_3 - E_2$ (d) $E_4 - E_3$

Fig. 5.19.: Reflectivity difference in dB (see color bar) of sensitivity results at an elevation of 2.5° (PPI mode) on 30 June 2012, 23:00 UTC

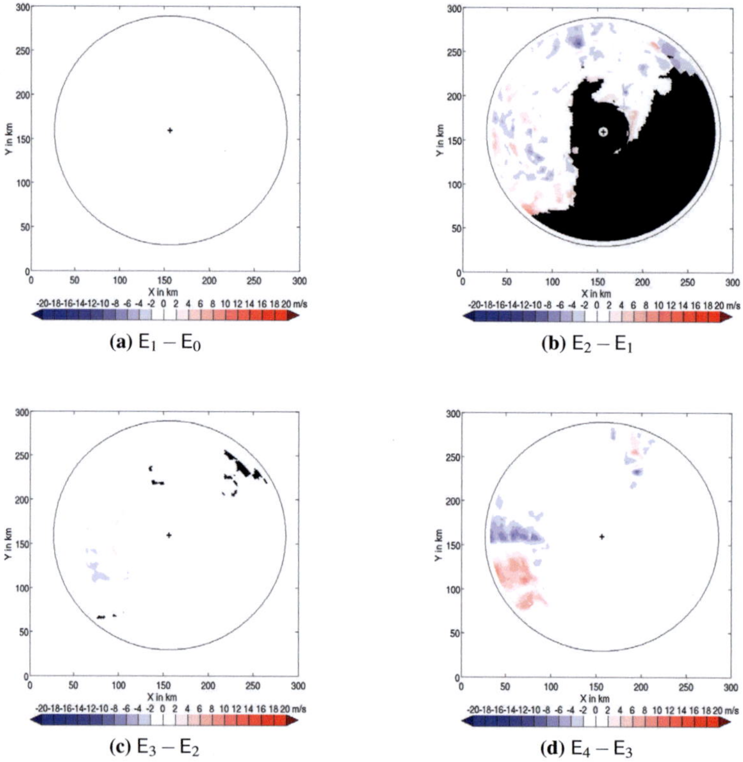

(a) $E_1 - E_0$ **(b)** $E_2 - E_1$

(c) $E_3 - E_2$ **(d)** $E_4 - E_3$

Fig. 5.20.: As Fig. 5.19 but for differences of Doppler velocity in m/s (see color bar)

By reason of elevations, thermodynamical profiles and model set-up, SODE results in a similar beam propagation as 43ERM. Pulse-volume averaging shows effects when radar beams arrive at long distances or intercept the melting level. The Mie-scattering scheme stimulates higher reflectivities than the Rayleigh method in case of large melting particles due to the special treatment of the refractive index by EMA. Taking attenuation into account has less significance in the stratiform event but has great impacts in convection. Weighting Doppler velocity by reflectivity allows for adjustment of areal coverage of Doppler velocity to the measured one.

With regard to the efficiency, the operator approaches a sufficient extent in all experiments, thanks to sophisticated vectorization and parallelization. The largest increase of computational time occurs when an intense pulse-volume averaging is applied. Another considerable increase is provoked by using the full Mie-scattering scheme instead of the Rayleigh approximation owing to estimation of numerous lookup tables. However, this work has to be done just once at the beginning of the model run and its computational significance will fade out as the model runs for a long period. Calculations of online beam propagation (SODE) and attenuation contribute slightly to the computational time.

Concerning the goal of experiments, which is to acquire a good balance between accuracy and efficiency of the operator, the question remains if such an intensive pulse-volume averaging is tolerable for purpose of data assimilation.

6. Data assimilation with Kalman Filter and its variants

Since Rudolf Kalman's first seminal paper on state estimation (Kalman, 1960), Kalman filter has been widely used in virtually many technical or quantitative fields such as fault detection, mathematical finance and global positioning, etc.. This recursive algorithm determines state variables of a noisy (linear) dynamical system by minimizing the analysis error of the current state in a root mean square sense as noisy measurements are taken and as the system propagates in time. Each update provides the latest unbiased state estimate together with a measure on the uncertainty of those estimates represented in form of a covariance matrix.

In this chapter, we begin with the mathematical and statistical background of a Kalman filter and then give an overview of derivation path to its variants. These information have been condensed out of a wealth of literature sources, mostly notably from Kalman (1960), Bouttier and Courtier (2002), Evensen (2003), Tippett et al. (2003), Wang et al. (2004) and Hunt et al. (2007).

6.1. Traditional Kalman Filter

A traditional Kalman Filter (hereafter KF) is a sequential method, which means that the model is integrated forward in time and before the next integration, the model state is reinitialized whenever observations are available. KF aims at finding an optimal analysis state $\mathbf{x}^a \in \mathbb{R}^n$ of the model, provided a forecast state $\mathbf{x}^f \in \mathbb{R}^n$ available at model grid points and a set of observations $\mathbf{y}^o \in \mathbb{R}^p$ available at irregularly distributed points, where p is the number of observations. Notice that a single vector of state $\mathbf{x}$ is formed by ordering the model variables by grid points and by variables, so the length n of $\mathbf{x}$ is the product of the number of grid points times the number of variables.

Let $\mathbf{x}_k^t \in \mathbb{R}^n$ be the unknown true state of the system at time t_k. We consider that the evolution of $\mathbf{x}_k^t$ is modeled by a discrete-time system:

$$\mathbf{x}_{k+1}^t = \mathscr{M}(\mathbf{x}_k^t) + \eta_k \quad , \tag{6.1}$$

where $\mathscr{M} : \mathbb{R}^n \longrightarrow \mathbb{R}^n$ is a nonlinear operator.

Eq. (6.1) defines thus the pseudo random model error by

$$\eta_k := \mathbf{x}_{k+1}^t - \mathscr{M}(\mathbf{x}_k^t) \quad , \tag{6.2}$$

with zero mean[1] and the corresponding error covariance matrix $\mathbf{Q}_k := \mathbf{E}\left(\eta_k \eta_k^T\right) \in \mathbb{R}^{n\times n}$, where $\mathbf{E}(\cdot)$ represents the statistical expected value.

Similarly, we define the forecast error by

$$\varepsilon_k^f := \underbrace{\mathscr{M}(\mathbf{x}_{k-1}^t) + \eta_{k-1}}_{=\mathbf{x}_k^t} - \underbrace{\mathscr{M}(\mathbf{x}_{k-1}^a)}_{=\mathbf{x}_k^f} \tag{6.3}$$

$$= \mathbf{x}_k^t - \mathbf{x}_k^f \quad , \tag{6.4}$$

with the corresponding forecast error covariance matrix $\mathbf{P}_k^f := \mathbf{E}\left(\varepsilon_k^f \varepsilon_k^{fT}\right) \in \mathbb{R}^{n\times n}$.

On the other hand, the relation between the true state and the observational variables is assumed to be described by the following expression:

$$\mathbf{y}_k^o = \mathscr{H}(\mathbf{x}_k^t) + \varepsilon_k^o \quad , \tag{6.5}$$

where $\mathscr{H} : \mathbb{R}^n \longrightarrow \mathbb{R}^p$ represents the observation forward operator that includes transformations from state variables into the observations and grid interpolations, and ε_k^o is the observation error with the corresponding covariance matrix $\mathbf{R}_k := \mathbf{E}\left(\varepsilon_k^o \varepsilon_k^{oT}\right) \in \mathbb{R}^{p\times p}$. Unfortunately, $\mathbf{R}_k$ is very hard to estimate and can cause problems in the analysis and quality control algorithms, therefore it makes sense in practice to try to minimize it by improving the accuracy of the model and forward operators, by using a bias correction scheme, by avoiding unnecessary observation preprocessing and by data thinning (Bouttier and Courtier, 2002). Several advanced techniques have been

[1] Although we assume that mean error is zero, in reality a model error has usually a bias needed to be taken into account.

developed for estimating $\mathbf{R}_k$ (Chapnik et al., 2004). However, at the early stage of this study, $\mathbf{R}_k$ is often assumed to be diagonal for simplicity, which means the observations are not correlated. This assumption applies also in this thesis.

Since we do not know exactly the true $\mathbf{x}_k^t$, we do not have full knowledge about the errors of forecast and observations, either. To deal with it, KF generally assumes a Gaussian forecast and observation error distribution, that is, $\mathfrak{p}(\varepsilon_k^f) = \mathscr{N}(0, \mathbf{P}_k^f)$ and $\mathfrak{p}(\varepsilon_k^o) = \mathscr{N}(0, \mathbf{R}_k)$, where $\mathfrak{p}$ is probability density function. In addition, we assume that the observation and forecast errors are uncorrelated:

$$\mathbf{E}\left(\varepsilon_k^o \varepsilon_k^{f^T}\right) = 0 \quad . \tag{6.6}$$

Furthermore, we define the analysis error by

$$\varepsilon_k^a = \mathbf{x}_k^a - \mathbf{x}_k^t \quad , \tag{6.7}$$

with the corresponding analysis error covariance matrix $\mathbf{P}_k^a := \mathbf{E}\left(\varepsilon_k^a \varepsilon_k^{a^T}\right) \in \mathbb{R}^{n \times n}$.

If we now assume that the dynamics and observations in Eqs. (6.1) and (6.5) are linear, the operators $\mathscr{M}$ and $\mathscr{H}$ will be replaced by two matrices $\mathbf{M}_k \in \mathbb{R}^{n \times n}$ and $\mathbf{H}_k \in \mathbb{R}^{p \times n}$, respectively, and we obtain:

$$\mathbf{x}_{k+1}^t = \mathbf{M}_k(\mathbf{x}_k^t) + \eta_k, \tag{6.8}$$

$$\mathbf{y}_k^o = \mathbf{H}_k(\mathbf{x}_k^t) + \varepsilon_k^o \quad . \tag{6.9}$$

For the "optimal" analysis, we aim to find the best estimates $\mathbf{x}_k^a$ of the state $\mathbf{x}_k^t$ using measurements $\mathbf{y}_k^o$. We say that $\mathbf{x}_k^a$ is optimal if the trace[2] of the analysis error covariance matrix $Tr(\varepsilon_k^a \varepsilon_k^{a^T})$ is minimized, and it is provided in the two following steps:

1.) Forecast step:

$$\mathbf{x}_k^f = \mathbf{M}_k \mathbf{x}_{k-1}^a \quad , \tag{6.10}$$

$$\mathbf{P}_k^f = \mathbf{M}_k \mathbf{P}_{k-1}^a \mathbf{M}_k^T + \mathbf{Q}_k \quad , \tag{6.11}$$

[2]In linear algebra, the trace of an $n \times n$ matrix A is defined as the sum of the elements on the main diagonal of A, i.e., $Tr(A) = a_{11} + a_{22} + \cdots a_{nn} = \sum_{i=1}^{n} a_{ii}$, where a_{ii} represents the entry on the $i-$th row and i-th column of A. The trace is invariant with respect to a change of basis.

2.) Analysis step:

$$\mathbf{K}_k = \mathbf{P}_k^f \mathbf{H}_k^T \left(\mathbf{H}_k \mathbf{P}_k^f \mathbf{H}_k^T + \mathbf{R}_k \right)^{-1} \quad , \tag{6.12}$$

$$\mathbf{P}_k^a = (\mathbf{I} - \mathbf{K}_k \mathbf{H}_k) \mathbf{P}_k^f (\mathbf{I} - \mathbf{K}_k \mathbf{H}_k) + \mathbf{K}_k \mathbf{R}_k \mathbf{K}_k^T = (\mathbf{I} - \mathbf{K}_k \mathbf{H}_k) \mathbf{P}_k^f \quad , \tag{6.13}$$

$$\mathbf{x}_k^a = \mathbf{x}_k^f + \mathbf{K}_k \left(\mathbf{y}_k^o - \mathbf{H}_k \mathbf{x}_k^f \right) \quad . \tag{6.14}$$

where the matrix $\mathbf{K}_k$ is the optimal weight matrix, also called Kalman gain.

$\mathbf{K}_k$ is obtained by the forecast error covariance in the observation space $\mathbf{P}_k^f \mathbf{H}_k^T$ multiplied by the inverse of the total error covariance (the sum of the forecast and observation error covariances) and can be intuitively understood to describe the correlation between state vector and observations. The relative magnitudes of matrices $\mathbf{R}_k$ and $\mathbf{H}_k \mathbf{P}_k^f \mathbf{H}_k^T$ control a relation between the filter's use of the forecast $\mathbf{x}_k^f$ and the observations $\mathbf{y}_k^o$: when the magnitude of $\mathbf{R}_k$ is small, meaning that the observations are accurate, the state estimate depends mostly on the observations; when the state is known accurately, then $\mathbf{H}_k \mathbf{P}_k^f \mathbf{H}_k^T$ is small compared to $\mathbf{R}_k$, and the filter mostly ignores the observations relying instead on the forecast. In Eq. (6.12), $\mathbf{K}_k$ linearly regresses the innovation (difference between the observation and the forecast in the observation space) onto state vector increments which are added to the forecast to generate the analysis. The analysis error covariance is given by the forecast error covariance multiplied by a matrix equal to the identity matrix minus the Kalman gain, and the data assimilation scheme uses observations to reduce forecast error covariance by the factor $(\mathbf{I} - \mathbf{K}_k \mathbf{H}_k)$.

KF is considered optimal when the following two assumptions are fulfilled:

1. Observation and forecast errors are Gauss-distributed and unbiased.

2. Observations are linearly related to the true model state (i.e., linear $\mathscr{H}$),

6.2. Extended Kalman Filter

Often in reality, $\mathscr{M}$ and $\mathscr{H}$ are nonlinear, so we have to consider approximate techniques for the non-linearity. One of the most widely used methods for state estimation of nonlinear system is the Extended Kalman Filter (EKF) that linearizes locally $\mathscr{M}$ and $\mathscr{H}$ (Bouttier and Courtier, 2002). If $\mathscr{M}$ is nonlinear, $\mathbf{M}_k$ can be defined as the tangent

linear operator (Jacobian) of $\mathscr{M}$ in the vicinity of $\mathbf{x}^a_{k-1}$, that is $\mathbf{M}_k := \left.\frac{\partial \mathscr{M}}{\partial \mathbf{x}}\right|_{\mathbf{x}=\mathbf{x}^a_{k-1}}$.
Similarly, $\mathscr{H}$ can be linearized in the vicinity of $\mathbf{x}^f_k$, which means $\mathbf{H}_k := \left.\frac{\partial \mathscr{H}}{\partial \mathbf{x}}\right|_{\mathbf{x}=\mathbf{x}^f_k}$.

6.3. Ensemble Kalman Filter

However, updating the forecast error covariance by Eq. (6.11) becomes very costly when dealing with complex data assimilation problems such as most meteorological and oceanographical models, owing to the massive dimensions of $\mathbf{M}$ and $\mathbf{P}^a$. In addition, the strongly nonlinear dynamics in these problems are difficult to linearize and the linearization generates instabilities which tend to make the filter diverge (Gauthier et al., 1993). Hence, it is necessary either to include empirical correction terms in the filter, or to use a more general stochastic forecast technique based on a Monte Carlo sampling of the filtering law, known as the Ensemble Kalman Filter (EnKF) (Evensen, 2003). EnKF represents the distribution of the system state using a collection of states, called an ensemble, and replace the covariance matrix by the sample covariance computed from this ensemble.

We start with an ensemble $\left\{\mathbf{x}^{a(i)}_{k-1} : i = 1,2,\ldots,N\right\}$ of model states at time t_{k-1}. The forecast step consists in evolving each ensemble member through the nonlinear dynamics (including the model noise simulation) to obtain a forecast ensemble at time t_k:

$$\mathbf{x}^{f(i)}_k = \mathscr{M}\left(\mathbf{x}^{a(i)}_{k-1}\right) + \eta^i_{k-1} \quad . \tag{6.15}$$

The empirical mean of the forecast ensemble is defined by

$$\overline{\mathbf{x}}^f_k = \frac{1}{N}\sum_{i=1}^{N} \mathbf{x}^{f(i)}_k \quad . \tag{6.16}$$

The empirical forecast ensemble error covariance matrix is then deduced from the following equation:

$$\mathbf{P}_k^f = \frac{1}{N-1}\sum_{i=1}^{N}\left(\mathbf{x}_k^{f(i)} - \overline{\mathbf{x}}_k^f\right)\left(\mathbf{x}_k^{f(i)} - \overline{\mathbf{x}}_k^f\right)^T = \frac{1}{N-1}\mathbf{X}_k^f\mathbf{X}_k^{f^T} \quad , \tag{6.17}$$

where

$$\mathbf{X}_k^f = \left[\mathbf{x}_k^{f(1)} - \overline{\mathbf{x}}_k^f, \mathbf{x}_k^{f(2)} - \overline{\mathbf{x}}_k^f, \ldots, \mathbf{x}_k^{f(N)} - \overline{\mathbf{x}}_k^f\right] \tag{6.18}$$

is the $n \times N$ matrix of forecast ensemble perturbations. Note that wherever an overbar is used in the context of a covariance estimate a factor of $N-1$ instead of N is implied in the denominator, so that the estimate is unbiased. $\mathbf{P}_k^f$ not just quantifies uncertainties of model forecasts and weights applied to the model with respect to observations, it also provides estimated correlations between variables of the model state for the propagation of the weighted information form the observed variables to the correlated ones, especially the unobserved ones. However, compared to Eq. (6.11), $\mathbf{P}_k^f$ in Eq. (6.17) can only represent the first term in Eq. (6.11) and hence is "blind" to model errors. A way to lighten this problem will be introduced later in Subsection 6.6.2.

The output of the analysis step is the analysis ensemble $\left\{\mathbf{x}_k^{a(i)} : i = 1, 2, \ldots, N\right\}$. The analysis mean is considered as the best estimate for the system state, defined by

$$\overline{\mathbf{x}}_k^a = \frac{1}{N}\sum_{i=1}^{N}\mathbf{x}_k^{a(i)} \quad , \tag{6.19}$$

with analysis ensemble error covariance matrix

$$\mathbf{P}_k^a = \frac{1}{N-1}\sum_{i=1}^{N}\left(\mathbf{x}_k^{a(i)} - \overline{\mathbf{x}}_k^a\right)\left(\mathbf{x}_k^{a(i)} - \overline{\mathbf{x}}_k^a\right)^T = \frac{1}{N-1}\mathbf{X}_k^a\mathbf{X}_k^{a^T} \quad , \tag{6.20}$$

where

$$\mathbf{X}_k^a = \left[\mathbf{x}_k^{a(1)} - \overline{\mathbf{x}}_k^a, \mathbf{x}_k^{a(2)} - \overline{\mathbf{x}}_k^a, \ldots, \mathbf{x}_k^{a(N)} - \overline{\mathbf{x}}_k^a\right] \tag{6.21}$$

is the $n \times N$ matrix of analysis ensemble perturbations.

The EnKF processes ensemble mean and each ensemble member using Eq. (6.14)

$$\overline{\mathbf{x}}_k^a = \overline{\mathbf{x}}_k^f + \mathbf{K}_k\left(\mathbf{y}_k^o - \mathbf{H}_k\overline{\mathbf{x}}_k^f\right) \quad , \tag{6.22}$$

$$\mathbf{x}_k^{a(i)} = \mathbf{x}_k^{f(i)} + \mathbf{K}_k\left(\mathbf{y}_k^o - \mathbf{H}_k\mathbf{x}_k^{f(i)}\right) \quad , \tag{6.23}$$

where $\mathbf{K}_k$ is given by Eq. (6.12) and $\mathbf{P}_k^f$ by Eq. (6.17). But it has been pointed out (Burgers et al., 1998) that a straightforward application of Eq. (6.23) to each ensemble member may cause an ensemble collapse, when the ensemble spread shrinks too rapidly. The traditional way of handling this problem proposed by Houtekamer and Mitchell (1998) and Burgers et al. (1998) is to update each ensemble member in a stochastic manner, instead of using a single realization of the observations $\mathbf{y}_k^o$, treating the observations as random variables by generating an ensemble of observations from a distribution with mean equal to the first-guess observation $\mathbf{y}_k^o$ and error covariance equal to $\mathbf{R}_k$, that is

$$\mathbf{R}_k = \mathbf{E}\left(\mathbf{Y}_k^o\mathbf{Y}_k^{o^T}\right) \quad , \tag{6.24}$$

yields

$$\mathbf{x}_k^{a(i)} = \mathbf{x}_k^{f(i)} + \mathbf{K}_k\left(\mathbf{y}_k^o + \varepsilon_k^{o(i)} - \mathbf{H}_k\mathbf{x}_k^{f(i)}\right) \quad , \tag{6.25}$$

where $\varepsilon_k^{o(i)}$ is a synthetic vector of perturbations of observations $\mathbf{y}_k^o$.

The ensemble average of Eq. (6.25) yields Eq. (6.22), provided that the ensemble average of $\varepsilon_k^{o(i)}$ is zero:

$$\mathbf{Y}_k^o\mathbf{1} = 0 \quad , \tag{6.26}$$

where $\mathbf{Y}_k^o = \left[\varepsilon_k^{o(i)}, \ldots, \varepsilon_k^{o(N)}\right]$ and $\mathbf{1} = [1, \ldots, 1]^T$. To compute the analysed error covariance, we first subtract the analysis Eq. (6.25) from (6.22) and gain the equation for the update of an ensemble perturbation:

$$\mathbf{X}_k^{a(i)} = \mathbf{X}_k^{f(i)} + \mathbf{K}_k \left(\varepsilon_k^{o(i)} - \mathbf{H}_k \mathbf{X}_k^{f(i)}\right) \quad , \tag{6.27}$$

which can be written in a matrix form for the full ensemble as

$$\mathbf{X}_k^a = \mathbf{X}_k^f + \mathbf{K}_k \left(\mathbf{Y}_k^o - \mathbf{H}_k \mathbf{X}_k^f\right) \quad . \tag{6.28}$$

The analyse error covariance carried by the ensemble can then be calculated using Eq. (6.20):

$$\begin{aligned}
\mathbf{P}_k^a &= \frac{1}{N-1}\left[\mathbf{X}_k^f + \mathbf{K}_k\left(\mathbf{Y}_k^o - \mathbf{H}_k\mathbf{X}_k^f\right)\right]\left[\mathbf{X}_k^f + \mathbf{K}_k\left(\mathbf{Y}_k^o - \mathbf{H}_k\mathbf{X}_k^f\right)\right]^T \\
&= \mathbf{P}_k^f - \mathbf{P}_k^f\mathbf{H}_k^T\mathbf{K}_k^T - \mathbf{K}_k\mathbf{H}_k\mathbf{P}_k^f + \mathbf{K}_k\mathbf{H}_k\mathbf{P}_k^f\mathbf{H}_k^T\mathbf{K}_k \\
&+ \frac{1}{N-1}\mathbf{K}_k\mathbf{Y}_k^o\mathbf{Y}_k^{o^T}\mathbf{K}_k^T + \frac{1}{N-1}\left(\mathbf{I} - \mathbf{K}_k\mathbf{H}_k\right)\mathbf{X}_k^f\mathbf{Y}_k^{o^T}\mathbf{K}_k^T \\
&+ \frac{1}{N-1}\mathbf{K}_k\mathbf{Y}_k^o\mathbf{X}_k^{f^T}\left(\mathbf{I} - \mathbf{H}_k^T\mathbf{K}_k^T\right) \quad .
\end{aligned} \tag{6.29}$$

In the EnKF, if all members are updated with the same observations without perturbations, that is, $\mathbf{Y}_k^o = 0$, the analysed error covariance produced by the ensemble becomes

$$\begin{aligned}
\mathbf{P}_k^a &= \mathbf{P}_k^f - \mathbf{P}_k^f\mathbf{H}_k^T\mathbf{K}_k^T - \mathbf{K}_k\mathbf{H}_k\mathbf{P}_k^f + \mathbf{K}_k\mathbf{H}_k\mathbf{P}_k^f\mathbf{H}_k^T\mathbf{K}_k \\
&= \left(\mathbf{I} - \mathbf{K}_k\mathbf{H}_k\right)\mathbf{P}_k^f\left(\mathbf{I} - \mathbf{H}_k^T\mathbf{K}_k^T\right) \quad .
\end{aligned} \tag{6.30}$$

Differing from Eq. (6.13), this expression contains one factor, $(\mathbf{I} - \mathbf{K}_k\mathbf{H}_k)$, too many. The missing term $\mathbf{K}_k\mathbf{R}_k\mathbf{K}_k^T$ causes $\mathbf{P}_k^a$ to be less than the value given by Eq. (6.13) and therefore results in underestimating analysis error and a premature reduction in the ensemble spread (Houtekamer and Mitchell, 1998; Burgers et al., 1998).

If there was $\mathbf{Y}_k^o$ that satisfies not just Eqs. (6.24) and (6.26) but also

$$\mathbf{X}_k^f\mathbf{Y}_k^{o^T}\mathbf{K}_k^T = 0 \quad , \tag{6.31}$$

then the analyzed error covariance (6.29) would exactly match the theoretical value (6.13). However, such a solution does not exist in general. Consequently, EnKF only satisfies these conditions approximately in a statistical sense, which can be expressed as (Burgers et al., 1998):

$$\mathbf{P}_k^a = (\mathbf{I} - \mathbf{K}_k\mathbf{H}_k)\mathbf{P}_k^f + O\left(\frac{1}{\sqrt{N}}\right) \quad . \tag{6.32}$$

This traditional algorithm is a stochastic filter and has become known as the perturbed observations EnKF. Its advantage is that the covariance matrices are no longer evolved using the forecast model like they are in KF, however, it introduces sampling errors which reduce the accuracy of the analysis covariance estimate, especially for small ensembles as Eq. (6.32) suggests. Therefore, designing an approach that does not require perturbed observations is desirable.

6.4. Ensemble Square Root Filter

Ensemble Square Root Filter (EnSRF) is a deterministic filter which means that no perturbed observations are used, and thus it differs from EnKF in the analysis step (Tippett et al., 2003).

We begin with substituting Eq. (6.17) to (6.12), which yields

$$\mathbf{K}_k = \frac{1}{N-1}\mathbf{X}_k^f\mathbf{X}_k^{f^T}\mathbf{H}_k^T\left(\frac{1}{N-1}\mathbf{H}_k\mathbf{X}_k^f\mathbf{X}_k^{f^T}\mathbf{H}_k^T + \mathbf{R}_k\right)^{-1} \quad . \tag{6.33}$$

At this point, we can effectively handle the nonlinear observation operator $\mathscr{H}$. In Eq. (6.33), we see that the linearized operator $\mathbf{H}_k$ always appears next to the matrix $\mathbf{X}_k^f$. The i-th column of $\mathbf{H}_k\mathbf{X}_k^f$ is $\mathbf{H}_k\left(\mathbf{x}^{f(i)} - \overline{\mathbf{x}}_k^f\right)$, which is the 1-st order Taylor approximation of $\mathscr{H}\left(\mathbf{x}^{f(i)}\right) - \mathscr{H}\left(\overline{\mathbf{x}}_k^f\right)$. Instead of linearizing $\mathscr{H}$ on the entire model space, we linearly approximate $\mathbf{H}_k\mathbf{X}_k^f$ by the $n \times N$ matrix of forecast ensemble observation perturbations

$$\mathbf{Y}_k^f = \left[\mathbf{y}_k^{f(1)} - \overline{\mathbf{y}}_k^f, \mathbf{y}_k^{f(2)} - \overline{\mathbf{y}}_k^f, \ldots, \mathbf{y}_k^{f(N)} - \overline{\mathbf{y}}_k^f\right] \quad , \tag{6.34}$$

where $\overline{\mathbf{y}}_k^f := \frac{1}{N}\sum_{i=1}^{N} \mathscr{H}\left(\mathbf{x}_k^{f(i)}\right)$ is the observation ensemble mean and $\mathbf{y}_k^{f(i)} := \mathscr{H}\left(\mathbf{x}^{f(i)}\right)$. Notice that the sum of the columns of $\mathbf{Y}_k^f$ is zero.

Consequently, we can rewrite Eq. (6.12) as

$$\mathbf{K}_k = \mathbf{X}_k^f \mathbf{Y}_k^{f^T} \underbrace{\left[\mathbf{Y}_k^f \mathbf{Y}_k^{f^T} + (N-1)\mathbf{R}_k\right]^{-1}}_{=:\mathbf{D}_k^{-1}} \quad , \tag{6.35}$$

where $\mathbf{D}_k \in \mathbb{R}^{N\times N}$ is positive definite and hence invertible.

The analysis ensemble mean is given by the following step:

$$\overline{\mathbf{x}}_k^a = \overline{\mathbf{x}}_k^f + \mathbf{K}_k \left(\mathbf{y}_k^o - \overline{\mathbf{y}}_k^f\right) \tag{6.36}$$

and the analysis covariance matrix can now be written as

$$\begin{aligned}
\mathbf{P}_k^a &= \frac{1}{N-1}\mathbf{X}_k^a \mathbf{X}_k^{a^T} \\
&= (\mathbf{I} - \mathbf{K}_k \mathbf{H}_k)\,\mathbf{P}_k^f \\
&= \left(\mathbf{I} - \mathbf{X}_k^f \mathbf{Y}_k^{f^T} \mathbf{D}_k^{-1} \mathbf{H}_k\right) \frac{1}{N-1}\mathbf{X}_k^f \mathbf{X}_k^{f^T} \\
&= \frac{1}{N-1}\mathbf{X}_k^f \left(\mathbf{I} - \mathbf{Y}_k^{f^T} \mathbf{D}_k^{-1} \mathbf{Y}_k^f\right) \mathbf{X}_k^{f^T} \quad ,
\end{aligned} \tag{6.37}$$

so the analysis ensemble perturbation matrix $\mathbf{X}_k^a$ is updated by

$$\mathbf{X}_k^a = \mathbf{X}_k^f \mathbf{W}_k^a \quad , \tag{6.38}$$

where $\mathbf{W}_k^a \in \mathbb{R}^{N\times N}$ is a matrix square root of $\mathbf{I} - \mathbf{Y}_k^{f^T} \mathbf{D}_k^{-1} \mathbf{Y}_k^f$ in the sense that

$$\mathbf{W}_k^{a^T} \mathbf{W}_k^a = \mathbf{I} - \mathbf{Y}_k^{f^T} \mathbf{D}_k^{-1} \mathbf{Y}_k^f \quad . \tag{6.39}$$

Thus the analysis states can now be updated as

$$\mathbf{x}_k^{a(i)} = \overline{\mathbf{x}}_k^a + \mathbf{X}_k^{f(i)} \quad . \tag{6.40}$$

Eq. (6.39) is essentially a Monte Carlo implementation of a square root filter (Maybeck, 1979), which explains the name of EnSRF. However, the matrix square roots in Eq. (6.39) are not unique, they can be computed in different ways, such as by Cholesky factorization or by singular value decomposition (Stoer, 1999), but they are all functionally equivalent, distinguishing only in algorithmic details. In the next section, one form of EnSRF is introduced, which computes the matrix square roots in the subspace spanned by the ensemble.

6.5. Ensemble Transform Kalman Filter

There are two difficulties in evaluating $\mathbf{D}_k^{-1}$. The first is size: a $p \times p$ matrix. For atmospheric applications where $p \approx O(10^5)$, it can be costly to gain the inverse of $\mathbf{D}_k$. The second difficulty occurs when the largest eigenvalue of $\mathbf{D}_k$ in Eq. (6.12) may be many orders of magnitude larger than its smallest eigenvalue. In this case, the matrix is ill-conditioned and hence very problematic to inverse. To implement it in an efficient manner, Ensemble Transform Kalman Filter (ETKF) rewrites the $\mathbf{K}_k$ so that the matrix inverse can be done in the ensemble space. ETKF makes use of the identity:

$$\mathbf{Y}_k^{f^T}\left[\mathbf{Y}_k^f\mathbf{Y}_k^{f^T}+(N-1)\mathbf{R}_k\right]^{-1}=\left[(N-1)\mathbf{I}+\mathbf{Y}_k^{f^T}\mathbf{R}_k^{-1}\mathbf{Y}_k^f\right]^{-1}\mathbf{Y}_k^{f^T}\mathbf{R}_k^{-1} \quad . \tag{6.41}$$

which can be verified if we multiply both sides of the equation on the right by $\left[\mathbf{Y}_k^f\mathbf{Y}_k^{f^T}+(N-1)\mathbf{R}_k\right]$ and consider that

$$\mathbf{Y}_k^{f^T}\mathbf{R}_k^{-1}\left[\mathbf{Y}_k^f\mathbf{Y}_k^{f^T}+(N-1)\mathbf{R}_k\right]=\left[(N-1)\mathbf{I}+\mathbf{Y}_k^{f^T}\mathbf{R}_k^{-1}\mathbf{Y}_k^f\right]\mathbf{Y}_k^{f^T} \quad ,$$

Now we have

$$\mathbf{K}_k=\mathbf{X}_k^f\left[(N-1)\mathbf{I}+\mathbf{Y}_k^{f^T}\mathbf{R}_k^{-1}\mathbf{Y}_k^f\right]^{-1}\mathbf{Y}_k^{f^T}\mathbf{R}_k^{-1} \quad , \tag{6.42}$$

and then Eq. (6.13) becomes

$$\begin{aligned}\mathbf{P}_k^a&=\frac{1}{N-1}\left(\mathbf{I}-\mathbf{K}_k\mathbf{H}_k\right)\mathbf{X}_k^f\mathbf{X}_k^{f^T}\\&=\frac{1}{N-1}\mathbf{X}_k^f\left\{\mathbf{I}-\frac{1}{N-1}\left(\mathbf{I}+\frac{1}{N-1}\mathbf{Y}_k^{f^T}\mathbf{R}_k^{-1}\mathbf{Y}_k^f\right)^{-1}\mathbf{Y}_k^{f^T}\mathbf{R}_k^{-1}\mathbf{Y}_k^f\right\}\mathbf{X}_k^{f^T} \quad .\end{aligned} \tag{6.43}$$

Applying the identity $\mathbf{I}-(\mathbf{I}+\mathbf{B})^{-1}\mathbf{B}=(\mathbf{I}+\mathbf{B})^{-1}$ with $\mathbf{B}=\frac{1}{N-1}\mathbf{Y}_k^{f^T}\mathbf{R}_k^{-1}\mathbf{Y}_k^f$, we can rewrite the analysis error covariance matrix as

$$\begin{aligned}
\mathbf{P}_k^a &= \frac{1}{N-1}\mathbf{X}_k^f\left(\mathbf{I}+\frac{1}{N-1}\mathbf{Y}_k^{f^T}\mathbf{R}_k^{-1}\mathbf{Y}_k^f\right)^{-1}\mathbf{X}_k^{f^T} \\
&= \mathbf{X}_k^f\underbrace{\left[(N-1)\mathbf{I}+\mathbf{Y}_k^{f^T}\mathbf{R}_k^{-1}\mathbf{Y}_k^f\right]^{-1}}_{=:\tilde{\mathbf{P}}_k^a}\mathbf{X}_k^{f^T} \\
&= \mathbf{X}_k^f\tilde{\mathbf{P}}_k^a\mathbf{X}_k^{f^T} \quad ,
\end{aligned} \tag{6.44}$$

where $\tilde{\mathbf{P}}_k^a\in\mathbb{R}^{N\times N}$ is the analysis error covariance matrix in the ensemble space that transforms the forecast ensemble perturbations into the analysis ensemble perturbations. Thus, we have

$$\begin{aligned}
\mathbf{X}_k^a &= \mathbf{X}_k^f\left[(N-1)\tilde{\mathbf{P}}_k^a\right]^{\frac{1}{2}} \\
&= \mathbf{X}_k^f\mathbf{W}_k^a
\end{aligned} \tag{6.45}$$

and

$$\mathbf{K}_k=\mathbf{X}_k^f\tilde{\mathbf{P}}_k^a\mathbf{Y}_k^f\mathbf{R}_k^{-1} \quad . \tag{6.46}$$

Instead of an expensive computation of $\mathbf{D}_k^{-1}$ in Eq. (6.35), Eq. (6.46) takes advantage of the fact that the matrix $\mathbf{R}_k$ is much easier to invert due to its typically diagonal or block diagonal structure and many or all of the blocks of $\mathbf{R}_k$ may remain the same from one analysis time to the next so $\mathbf{R}_k^{-1}$ need not be recomputed each time. In addition, the matrix inverse $\tilde{\mathbf{P}}_k^a$ is done within the ensemble space, which usually has a much smaller dimension than the observation space ($N\ll p$).

Therefore, Eq. (6.36) becomes

$$\begin{aligned}
\overline{\mathbf{x}}_k^a &= \overline{\mathbf{x}}_k^f+\mathbf{X}_k^f\underbrace{\tilde{\mathbf{P}}_k^a\mathbf{Y}_k^f\mathbf{R}_k^{-1}\left(\mathbf{y}_k^o-\overline{\mathbf{y}}_k^f\right)}_{:=\overline{\mathbf{w}}_k^a} \\
&= \overline{\mathbf{x}}_k^f+\mathbf{X}_k^f\overline{\mathbf{w}}_k^a \quad ,
\end{aligned} \tag{6.47}$$

where $\overline{\mathbf{w}}_k^a$ is the analysis increment in the ensemble space.

In Eq. (6.45), $\mathbf{W}_k^a$ is the symmetric square root of the analysis error covariance matrix in ensemble space. Since $\tilde{\mathbf{P}}_k^a$ is symmetric positive definite matrix, it always has a unique symmetric positive definite square root (Halmos, 1974). It is numerically solved by singular vector decomposition (SVD) using eigenvalues and eigenvectors of $(N-1)\tilde{\mathbf{P}}_k^a$:

$$\mathbf{W}_k^a = U\Sigma^{\frac{1}{2}}U^T \quad , \tag{6.48}$$

where U is an orthogonal matrix of eigenvectors of $(N-1)\tilde{\mathbf{P}}_k^a$ and $\Sigma^{\frac{1}{2}}$ is the diagonal matrix of the square roots of the eigenvalues.

With this choice for $\mathbf{W}_k^a$, we obtain

$$\begin{aligned} \mathbf{P}_k^a &= \mathbf{X}_k^f \tilde{\mathbf{P}}_k^a \mathbf{X}_k^{f^T} \\ &= \mathbf{X}_k^f \frac{1}{N-1} \mathbf{W}_k^a \mathbf{W}_k^{a^T} \mathbf{X}_k^{f^T} \\ &= \frac{1}{N-1} \mathbf{X}_k^a \mathbf{X}_k^{a^T} \end{aligned} \tag{6.49}$$

and as shown in Wang et al. (2004) that the sum of columns $\mathbf{X}_k^{a(i)}$ of $\mathbf{X}_k^a$ is zero, so the analysis ensemble has the correct sample mean.

Another reason in favor of the use of the symmetric square root to calculate $\mathbf{W}_k^a$ from $\tilde{\mathbf{P}}_k^a$ is that it ensures the continuous dependency of $\mathbf{W}_k^a$ on $\tilde{\mathbf{P}}_k^a$ (Hunt et al., 2007), which is substantial in a local analysis scheme, as we will see in the next section, so that adjacent analysis points, whose corresponding local forecast ensemble has small disparities, will differ slightly in $\tilde{\mathbf{P}}_k^a$. The derived symmetric square root matrix can carry such characteristics and result in similar analysis ensemble perturbations at adjacent points and thus smoothness in the analysis. Another desirable property of the symmetric square root is that it minimize the distance[3] between $\mathbf{W}_k^a$ and the identity matrix, thus the analysis ensemble perturbations are in this sense as close as possible to the the forecast ensemble perturbations subject to the constraint on the analysis error covariance matrix (Ott et al., 2004). Harlim (2006) showed that the symmetric solution outperforms a non-symmetric one, given the same ensemble size.

[3]The distance is measured by the Frobenius norm.

Once $\overline{\mathbf{x}}_k^a$ and $\mathbf{P}_k^a$ are specified, we have to construct an analysis ensemble of model states, whose mean is $\overline{\mathbf{x}}_k^a$ and whose error covariance matrix satisfies $\mathbf{P}_k^a = \frac{1}{N-1}\mathbf{X}_k^a\mathbf{X}_k^{a^T}$. To fulfill these conditions, we generate the ensemble using

$$\begin{aligned}
\mathbf{x}_k^{a(i)} &= \overline{\mathbf{x}}_k^a + \mathbf{X}_k^f \mathbf{W}_k^{a(i)} \\
&= \overline{\mathbf{x}}_k^f + \mathbf{X}_k^f \overline{\mathbf{w}}_k^a + \mathbf{X}_k^f \mathbf{W}_k^{a(i)} \\
&= \overline{\mathbf{x}}_k^f + \mathbf{X}_k^f \left(\overline{\mathbf{w}}_k^a + \mathbf{W}_k^{a(i)} \right) \quad ,
\end{aligned} \tag{6.50}$$

where $\mathbf{W}_k^{a(i)}$ is the ith column of symmetric square root matrix $\mathbf{W}_k^a = \left[(N-1)\tilde{\mathbf{P}}_k^a\right]^{\frac{1}{2}}$.

6.6. Local Ensemble Transform Kalman Filter

Hunt et al. (2007) adopted an alternative algorithm of ETKF, named Local Ensemble Transform Kalman Filter (LETKF), by performing the analyses locally in space. The fundamental difference between LETKF and ETFK is, as the names suggest, the localization, which ameliorates computational efficiency because the analyses at different model grid points are independent and can be done in parallel. Furthermore, since observations are assimilated simultaneously, not serially, it is simple to take observation error correlations into account. In what follows, we will introduce the corresponding techniques including localization and covariance inflation.

6.6.1. Localization

In fact, the forecast covariance matrix $\mathbf{P}_k^f$ in all previous mentioned forms of Ensemble Kalman filters suffer from a sampling error which increases as the absolute value of the correlation between an observation and a state variable becomes weak. To estimate those weak correlations precisely, thousands of ensemble members are required in any case, thus for the limited size of ensemble in practical use we have to somehow deal with the sampling error associated with weak correlations. Although the expected correlation is usually not known a priori, it is generally believed that the correlation weakens with physical distances between an observations and a state variable. As observations become remote from the state variables, their potential positive impact can be expected to be insignificant. So we consider only the observations from a local

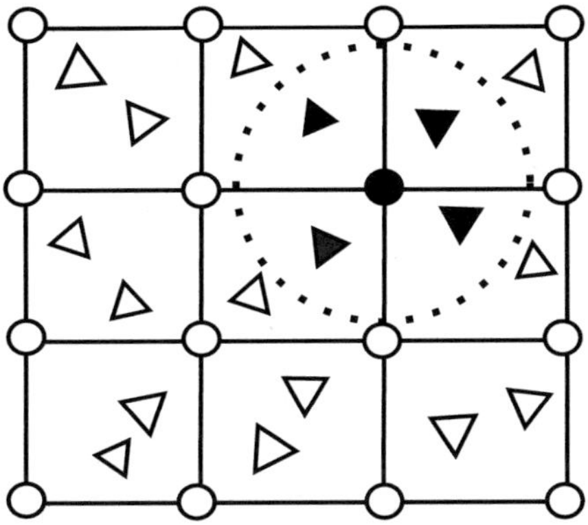

Fig. 6.1.: Top view of a localized domain: model grid and irregularly distributed observations are denoted with ○ and △, respectively, while the current analysis grid point and its associated observations are found within the localization radius (indicated by the dashed line).

domain surrounding the location of the analysis (Keppenne, 2000; Ott et al., 2002, 2004) and multiply the entries in $\mathbf{R}_k^{-1}$ by a factor that decays from one to zero as the distance of the observations from the analysis grid point increases, which corresponds to gradually increasing the uncertainty assigned to the observations until beyond a certain distance they have infinite uncertainty and thus no influence on the analysis. The choice of the size of the local domains should reflect the distance over which dynamical correlations represented by the ensemble are meaningful. A common idea is to use observations within a cylinder of a given radius and height centered at the analysis grid point and to determine empirically which value of volume produces the best results (see Fig. 6.1). This is denoted as explicit localization. Another localization is done implicitly by multiplying the elements in $\mathbf{P}_k^f$ with a distance-depend weight function. If the expected absolute value of correlation is small enough, the weight can be equal to zero and the regression does not need to be done. The most popular weight function $\chi(d,c)$ is given by Gaspari and Cohn (1999), where d is the distance and c half-width. For $d \geq 2c$, the observation has no impact on the state variable; for $d < 2c$, χ behaviors like a Gaussian. The optimal value of c depends on the size of the ensemble, with increasing size, corrections at larger and larger distances can be precisely estimated. Therefore, an appropriate c can be specified for a certain ensemble size. In this work, the explicit manner is employed in which we choose a local subset of the global observations within a local domain around each grid point of the model and conduct separate analyses simultaneously using only the local observations.

A complementary reason favoring the localization is deduced from the fact that the rank of $\mathbf{P}_k^f$ is equal to the rank of $\mathbf{X}_k^f$, which is at most $N-1$ because the sum of its columns is equal to 0. Consequently, the ensemble size limits the rank of $\mathbf{P}_k^f$ that can represent uncertainty only in $(N-1)$-dimensional subspace spanned by the columns of $\mathbf{X}_k^f$ and a global analysis will allow correction to the model state only in this subspace, which means that forecast errors will increase in directions not included in this space and will not be adjusted by the analysis step. If the observations are plentiful whereas the ensemble size N is small, the analysis will inevitably smooth the observational information, which leads to a loss of analysis accuracy and divergence from the real state. Consequently, we need to make the ensemble size commensurable with the number of observed degrees within the global model to provide an accurate representation, which makes the algorithm computationally very inefficient. As recognized in Oczkowski et al. (2005) and Patil et al. (2001), the smaller the local domain in a model we choose, the smaller the ensemble size is necessary to properly represent the model dynamics in the local domain. So if we carry out the analysis step locally by choosing different linear combinations of the ensemble members in different domains, the selected ensembles need to represent uncertainty only in the local domain and the rank problem is mitigated to a high extent. The global analysis that becomes a larger amount of small local problems, each of which individually has no rank problem, is able to fit much higher observed degrees of freedom (Fukumori, 2002; Ott et al., 2002, 2004).

6.6.2. Covariance inflation

Besides sampling errors, ensemble Kalman filters are also subject to other sources of errors, such as model errors and interpolation as well as representativeness errors of the operator. All these errors can potentially cause underestimation of the forecast covariance and overconfidence in the forecast state estimate. As previously stated, the ensemble spread $\mathbf{P}_k^f$ ignores model errors. Moreover, when the observations are dense, $\mathbf{P}_k^f$ will be reduced massively and turns out to be too small. As spread represents the uncertainty, the filter believes that it performs better than it does in reality and thus the analysis loses track of the truth. In order to compensate for this tendency, an ad hoc approach is commonly employed which artificially inflates the forecast error covariance matrix $\mathbf{P}_k^f$ before each analysis (alternatively, one could inflate the analysis error covariance matrix $\mathbf{P}_k^a$ after each analysis). Although the standard covariance inflation method is to multiply the forecast ensemble perturbations $\mathbf{X}_k^f$ by an appropriate constant factor $\sqrt{\beta} > 1$, which is equal to multiplying $\mathbf{P}_k^f$ by β, one can also attain similar results in a more efficient way which leaves $\mathbf{X}_k^f$ alone but rewrites $\tilde{\mathbf{P}}_k^a$ as follows

$$\tilde{\mathbf{P}}_k^a = \left[(N-1)\mathbf{I}/\beta + \mathbf{Y}_k^{f^T}\mathbf{R}_k^{-1}\mathbf{Y}_k^f\right]^{-1} \quad . \tag{6.51}$$

It can be shown that this modification has the same effect on the analysis mean $\mathbf{x}_k^{a(i)}$ and covariance matrix $\mathbf{P}_k^a$ as multiplying $\mathbf{X}_k^f$ and $\mathbf{Y}_k^f$ by $\sqrt{\beta}$, respectively. Notice that in case of a linear $\mathscr{H}$, this is same as inflating the forecast ensemble by $\sqrt{\beta}$ before applying $\mathscr{H}$ to get $\mathbf{Y}_k^f$. If β is close to one, this is a good approximation even for a nonlinear $\mathscr{H}$.

On the other hand, covariance inflation can be thought as applying a damping factor to the influence of previous observations on the current analysis. Because this damping factor is applied in each analysis step, the cumulative effect is to diminish the influence of an observation on future analyses exponentially with time, so the inflation factor determines the time scale over which observations have an influence. This effect is particularly advantageous in the presence of model errors, because then the model can only reliably propagate information given by the observations for a limited time range. In this sense, covariance inflation also localizes the analysis in time.

But a single value of inflation is not appropriate for all state variables since the ensemble spread is very sensitive to the observing density. When the observations are dense, the ensemble spread is cut down excessively, which means that the model is too confident, so a larger value of β should be taken, and vice versa. A fixed β can result in values that are inconsistent with climatological values, and in the worse case, incompatible with model's numerical methods (Anderson, 2008), leading to model failure. So an adaptive inflation is preferable (Li et al., 2009), which proposed an online estimation of the inflation factor. The idea is to compare the "observed" observation-minus-forecast, given by $[\mathbf{y}^o - \mathscr{H}(\mathbf{x}^f)]$, with the "predicted" one, given by $(\mathbf{R}+\mathbf{HP}^f\mathbf{H}^T)$. This method was adopted in a LETKF environment by Bonavita et al. (2010), where β was time and space dependent.

6.6.3. Implementation of LETKF

In this section, we explain how the LETKF algorithm given above is implemented. This method can be summarized by ten steps described below. The inputs to the analysis are a forecast ensemble of $n_{[g]}$-dimensional model state vectors $\left\{\mathbf{x}_{[g]}^{f(i)} : i=1,2,\ldots,N\right\}$, a nonlinear operator $\mathscr{H}_{[g]}$ from the $n_{[g]}$-dimensional model space to the $p_{[g]}$-dimensional observation space, an $p_{[g]}$-dimensional vector $\mathbf{y}_{[g]}^o$ of observations, and an $p_{[g]} \times p_{[g]}$ observation error covariance matrix $\mathbf{R}_{[g]}$. The subscript $[g]$ refers to the global model state and all available observations, from which a local subset will be chosen for each local analysis and the subscript $[l]$ reflects a local domain associated with an arbitrary grid point. Step 1 and 2 are basically global operations, but can be carried out locally in a parallel scheme, if $\mathscr{H}$ is a local interpolation operator. In Step 3, for each model grid point, we truncate $\overline{\mathbf{x}}_{[g]}^f$ and $\mathbf{X}_{[g]}^f$ to contain only the model variables for that grid point, and truncate $\mathbf{y}_{[g]}^o$, $\overline{\mathbf{y}}_{[g]}^f$ and $\mathbf{Y}_{[g]}^f$ to contain observations within a local domain around that point. After the local analysis for each grid point is done separately in Step 3 - 8, the final result of the global analysis is given in Step 9. In Step 10, the new global forecast ensemble is created.

In details the steps are:

1. Apply the operator $\mathscr{H}_{[g]}$ to each $\mathbf{x}_{[g]}^{f(i)}$ to obtain the global forecast observation ensemble $\left\{\mathbf{y}_{[g]}^{f(i)}\right\}$, and calculate the ensemble average $\overline{\mathbf{y}}_{[g]}^f$ of $\left\{\mathbf{y}_{[g]}^{f(i)}\right\}$. Construct the global forecast observation ensemble perturbation $p_{[g]} \times N$ matrix $\mathbf{Y}_{[g]}^f$ by taking its columns to be the vectors obtained by substracting $\overline{\mathbf{y}}_{[g]}^f$ from each $\mathbf{y}_{[g]}^{f(i)}$.

2. Calculate the ensemble mean $\overline{\mathbf{x}}_{[g]}^f$ of $\left\{\mathbf{x}_{[g]}^{f(i)}\right\}$ and subtract $\overline{\mathbf{x}}_{[g]}^f$ from each $\mathbf{x}_{[g]}^{f(i)}$ to build the columns of global forecast ensemble perturbation $n_{[g]} \times N$ matrix $\mathbf{X}_{[g]}^f$.

3. Select all necessary data needed to obtain the analysis ensemble at a given grid point. Select the rows of $\overline{\mathbf{x}}_{[g]}^f$ and $\mathbf{X}_{[g]}^f$ corresponding to the given grid point, forming their local counterparts: the $n_{[l]}$-dimensional vector $\overline{\mathbf{x}}_{[l]}^f$ and the $n_{[l]} \times N$ matrix $\mathbf{X}_{[l]}^f$. Select the rows of $\overline{\mathbf{y}}_{[g]}^f$ and $\mathbf{Y}_{[g]}^f$ corresponding to the observations

chosen for the analysis in the local domain to form the $p_{[l]}$-dimensional vector $\overline{\mathbf{y}}^f_{[l]}$ and $p_{[l]} \times N$ matrix $\mathbf{Y}^f_{[l]}$. Select the corresponding rows of $\mathbf{y}^o_{[g]}$ to form the $p_{[l]}$-dimensional vector $\overline{\mathbf{y}}^o_{[l]}$. Select the corresponding rows and columns of $\mathbf{R}_{[g]}$ to form $p_{[l]} \times p_{[l]}$ matrix $\mathbf{R}_{[l]}$.

4. Calculate the $N \times p_{[l]}$ matrix $\mathbf{C}_{[l]} = \mathbf{Y}^{fT}_{[l]} \mathbf{R}^{-1}_{[l]}$. Since this is the only step where $\mathbf{R}_{[l]}$ is used, it is much more convenient to estimate $\mathbf{C}_{[l]}$ by solving the linear system $\mathbf{R}_{[l]} \mathbf{C}^T_{[l]} = \mathbf{Y}^f$ than inverting $\mathbf{R}_{[l]}$.

5. Calculate the $N \times N$ matrix $\tilde{\mathbf{P}}^a_{[l]} = \left[(N-1)\mathbf{I}/\beta + \mathbf{Y}^{fT}_{[l]} \mathbf{R}^{-1}_{[l]} \mathbf{Y}^f_{[l]}\right]^{-1}$.

6. Calculate the $N \times N$ matrix $\mathbf{W}^a_{[l]} = \left((N-1)\tilde{\mathbf{P}}^a_{[l]}\right)^{\frac{1}{2}}$.

7. Calculate the N-dimensional vector $\overline{\mathbf{w}}^a_{[l]} = \tilde{\mathbf{P}}^a_{[l]} \mathbf{C}_{[l]} \left(\mathbf{y}^o_{[l]} - \overline{\mathbf{y}}^o_{[l]}\right)$ and add it to each column of $\mathbf{W}^a_{[l]}$, The columns of resulting $N \times N$ matrix are the weight vectors $\left\{\mathbf{w}^{a(i)}\right\}$.

8. Calculate the analysis ensemble members at the analysis grid point by $\mathbf{x}^{a(i)}_{[l]} = \mathbf{X}^f_{[l]} \overline{\mathbf{w}}^a_{[l]} + \mathbf{x}^{f(i)}_{[l]}$.

9. After completing Step 3-8 for each grid point, the results of Step 8 are gathered to form the global analysis ensemble $\left\{\mathbf{x}^{a(i)}_{[g]}\right\}$.

10. Calculate the new global forecast ensemble $\left\{\mathbf{x}^{f(i)}_{[g]}\right\}$ by $\mathbf{x}^{f(i)}_{[g]} = \mathscr{M}\left(\mathbf{x}^{a(i)}_{[g]}\right)$.

6.7. 4-Dimensional Local Ensemble Transform Kalman Filter

In an operational setting, the analyses are generated at several hours intervals, though many observations are available more frequently. Since significant changes could happen over such a time interval, it is reasonable to consider observations at intermediate times than to pretend that they were taken at the analysis time. Hunt et al. (2004) extended the LETKF to a four-dimensional version 4D Local Ensemble Transform Kalman Filter (4D-LETKF) which estimates the analysis ensemble mean by fitting the linear combinations of the trajectories of the background ensemble to all of the observations collected between two analysis times.

Recall that in Section 6.5 we wrote the analysis mean as $\overline{\mathbf{x}}^a = \overline{\mathbf{x}}^f + \mathbf{X}^f \overline{\mathbf{w}}^a$ (cf. Eq. (6.47), the subscript k is omitted here and hereafter for brevity),where $\overline{\mathbf{w}}^a$ is determined by $\mathbf{R}$, $\mathbf{y}^o$, $\overline{\mathbf{y}}^f$ and $\mathbf{Y}^f$. Essentially, $\overline{\mathbf{w}}^a$ specifies the linear combination of background ensemble states that best fit $\mathbf{y}^o$. Moreover, $\mathbf{y}^o$ and $\mathbf{Y}^f$ are formed by mapping the background ensemble into the observational space. So for observations taken at different times $\mathbf{y}^o$ and $\mathbf{Y}^f$ must be accordingly redefined. To be more concrete, let's assume that we have data $(t_j, \mathbf{y}^o_{t_j})$ from various times t_j since the last analysis. Let $\overline{\mathbf{x}}^f_{t_j}$ and $\mathbf{X}^f_{t_j}$ be the ensemble background mean and matrix of background ensemble perturbations at time t_j. Let H_{t_j} be the observation operator for time t_j and $\mathbf{R}_{t_j}$ the error covariance matrix for these observations. So now, for each t_j, we apply H_{t_j} to the background ensemble state $\mathbf{x}^{f(i)}_{t_j}$ to gain vectors $\mathbf{y}^{f(i)}_{t_j}$, average those vectors to gain $\overline{\mathbf{y}}^f_{t_j}$, and subtract $\overline{\mathbf{y}}^f_{t_j}$ from $\mathbf{y}^{f(i)}_{t_j}$ to get the columns of $\mathbf{Y}^f_{t_j}$. Then, a combined observation vector $\mathbf{y}^o$ is formed by vertically concatenating the column vectors $\mathbf{y}^o_{t_j}$, and analogously by vertical concatenation of the vectors $\overline{\mathbf{y}}^f_{t_j}$ and matrices $\mathbf{Y}^f_{t_j}$, respectively, we build the combined background observation mean $\overline{\mathbf{y}}^f$ and perturbation matrix $\mathbf{Y}^f$. The corresponding $\mathbf{R}$ is a block diagonal matrix with blocks $\mathbf{R}_{t_j}$ (this assumes that observations taken at different times have uncorrelated errors, though such correlations if present could be included in $\mathbf{R}$).

Accordingly, the steps in Subsection 6.6.3 should be modified. Step 1 is executed for each observation time t_j and the results are combined as described in the previous paragraph to form $\overline{\mathbf{y}}^f$ and $\mathbf{Y}^f$. However, Step 2 is only carried out only at the analysis time and save the resulting $\overline{\mathbf{x}}^f$ and $\mathbf{X}^f$ for use in Step 8.

6.8. Data thinning

Although the high frequency of radar observations is very advantageous for estimating an initial state for NWP, a huge amount of assimilated observations may also lead to high computational costs, massive memory space allocation and very time-consuming data transmission. Additionally, dense data can severely violate the assumption of independent observation errors made in most assimilation schemes, including Kalman filters. The error correlations are unknown a priori, and calculations of these correlations in the assimilation system would require more complex observation error statistics and evoke increased computational costs. To combat these problems, efforts must be made to cut back on the amount of observations before assimilation, meanwhile, the quality of assimilation should be preserved or even be enhanced. Since error correlations often exist for observations lying close together, if we dilute the observations so sparse that the distances among observations are larger than the correlation length, the observations can be considered as uncorrelated. Currently, a simple data thinning method is implemented in the operator, where observations can be thinned in radial distance and azimuthal scanning angle at specified intervals. The crudeness of this technique is that the problem of inhomogeneous radar data distribution (more data at closer distance from radar) is not alleviated by that.

6.9. Statistics used for verification of assimilation performance

A fundamental assumption of the standard KF is that the observations and the model outputs are unbiased. An observation bias typically indicates instrumental inaccuracies, representativeness errors, or, in the case of remote sensing observations, errors in the retrieval algorithm. After quality control the observations can be often but not always assumed to be unbiased (Lorenc and Hammon, 1988). In contrast, model forecasts are hardly ever unbiased. Model forecast errors depend on model structure, parameters, discretization as well as model initial conditions. Generally, the forecast error contains a random and a systematic component. The latter one is referred to as model bias.

Let $\mathbf{y}^o = [y_1^o, \dots, y_p^o]$ be set of observations. We define bias of forecast ensemble mean by

$$Bias^f = \frac{1}{p} \sum_{i=1}^{p} \left[y_i^o - \bar{y}_i^f \right] \quad . \tag{6.52}$$

It is the difference between the observation and the forecast projected to the observation points. $Bias^f$ is positive when the model overestimates the observations, while a negative bias reveals underestimation of the observations.

A standard measure for the misfit of simulated ensemble mean and observations is the root mean square error (RMSE)

$$RMSE^f = \sqrt{\frac{1}{p-1} \sum_{i=1}^{p} \left[y_i^o - \bar{y}_i^f \right]^2} \quad , \tag{6.53}$$

which is considered as the expected spread. In contrast to $Bias^f$, $RMSE^f$ quantifies the error of the estimates in a least-square sense and it is always larger than or equal to zero. A small $RMSE^f$ corresponds to a good fit.

To estimate the reliability of assimilation results (Sacher and Bartello, 2009), $RMSE^f$ is usually compared with the aforementioned spread, which is considered as a measure for the ensemble variability, defined as

$$\begin{aligned} Spread^f &= \sqrt{\frac{Spread_1^{f^2} + Spread_2^{f^2} + \cdots + Spread_p^{f^2}}{p-1}} \\ &= \sqrt{\frac{1}{p-1}\sum_{i=1}^{p} Spread_i^{f^2}} \quad , \end{aligned} \tag{6.54}$$

where

$$Spread_i^f = \sqrt{\frac{(y_i^{f(1)} - \bar{y}_i^f)^2 + (y_i^{f(2)} - \bar{y}_i^f)^2 + \cdots + (y_i^{f(N)} - \bar{y}_i^f)^2}{N-1}}, i = 1, 2, \ldots, p \tag{6.55}$$

is root mean square difference between the forecast ensemble members and the forecast ensemble mean. $Spread^f$ can be interpreted as the assumed forecast error covariance.

In analogy to definitions of $Bias^f$, $RMSE^f$ and $Spread^f$ before the analysis step, we can also define $Bias^a$, $RMSE^a$ and $Spread^a$ after the analysis step. All of them are used to quantify the assimilation performance and to represent the model/observations consistency. If RMSE and spread are approximately identical, the ensemble variance captures the estimation error correctly and thus the ensemble forecasting system is reliable. In practice, the RMSE is always higher than the actual spread and an increase in the spread usually helps to reduce the RMSE and give a better performance of the data assimilation system.

7. Data assimilation experiments using 4D-LETKF for the convective event of 31 May 2011

In this chapter, first experiments applying the new radar forward operator in the data assimilation procedure are performed. The used data assimilation scheme is a newly developed 4D-LETKF software package from DWD (Schraff et al., 2012).

Generally, the best way to examine a newly built data assimilation system and to investigate the potential impacts of new observations on it is to conduct Observing System Simulation Experiments (OSSEs, Arnold Jr. and Dey (1986)). The OSSE methodology begins with a free running mode without data assimilation to provide the "truth" and generate simulated observations with realistic errors. This run is called Nature Run (NR). Then, two experiments are to be carried out: a control run, in which current observational data are included, used to generate a reference field and a perturbation run, in which simulated new observations under evaluation are added. By comparison of results of these two runs, we are able to evaluate beforehand the improvement in forecast skill due to the proposed new data.

There are several advantages of OSSEs, such as easy control of the experiments, precise knowledge of the data properties and errors, and knowledge of the truth. However, OSSEs require immense resources in maintenance and computing power (McCarty, 2012), particularly when radar data are involved. Unfortunately, a first version of the 4D-LETKF data assimilation system at DWD has been just finished at the time of this writing, such that, among the other ongoing work, a framework for OSSEs is still under construction at present. Therefore, we have to apply the data assimilation system to a real case, for which the characteristics of model and observation errors are not known precisely. Considering the assimilation of radar reflectivity would be more problematic than Doppler velocity for several reasons:

1. Model errors are expected to be particularly large for reflectivity (Gilmore et al., 2004). Whereas simulated Doppler velocity is computed from the 3D wind vector, which are controlled by grid-scale dynamics, simulated reflectivity is computed

from hydrometeor fields, which are controlled by microphysical parametrizations and their inherent uncertainties.

2. Even if the hydrometeor fields are predicted well by the model, there are considerable uncertainties in how to calculate reflectivity from model hydrometeor fields. particularly for ice and mixed-phase precipitation, i.e., bias errors in the operator,

3. Reflectivity is strongly nonlinearly tied to model variables (Tong and Xue, 2005),

4. Unlike Doppler velocity observation errors, reflectivity observation errors are affected significantly by attenuation and errors in radar calibration (Wilson and Brandes, 1979).

Therefore, we prefer assimilation of Doppler velocity as a start point to collect first experiences, and assimilation of reflectivity will be investigated in future.

Additionally, a verification tool for 4D-LETKF results (typically, analysis mean against observations in terms of surface precipitation) and an algorithm that combines the post-processing quality products with radar observations to filter contaminated data are currently also under development. Constrained by these conditions, the experiments results have to be taken only as first steps and a technical proof of concept. More work will be done in the future to tune the system towards operational application.

7.1. Description of weather conditions, model and 4D-LETKF setup

On the morning of 31 May 2011, the cold front of low pressure system "Yves" arrived in West Germany and during day advanced slowly eastwards. Due to the extreme temperature changes across the front boundary (hardly more than $13°C$ in the west, 25 - $32°C$ in the east), severe thunderstorms occurred along the frontal zone. After the front passage, the thunderstorms partly turned into persistent rain (see Fig. 7.1).

All data assimilation experiments described in detail in the next section are run on 8 processors of the NEC SX-9 cluster at DWD. The COSMO-DE-model is operated with an horizontal resolution of $(0.025°, 0.025°)$ and 51 vertical levels within the domain depicted in Fig. 2.1b. An ensemble size of at least 40, suggested by Chris Snyder at the COSMO GM at Athens, is expected to be sufficient to render good results. To create the ensemble members, we start with 40 identical members and add random perturbations coming from the boundary data of GME-LETKF. These perturbations will be transported into the internal domain by the model itself[1]. Starting at 18:00 UTC 31 May 2011, we perform a 1-hour data assimilation cycle until 21:00 UTC, i.e., the observations are assimilated every one hour. It is assumed throughout that the observation error has a Gaussian distribution with 1 m/s for Doppler velocity. To get rid of outlying observations, especially in the radar data, a rough solution, though not necessarily a bad one, is to reject all data that have a potential of being contaminated, i.e., it is claimed that

$$|\mathbf{y}^o - \overline{\mathbf{y}}^f| > 3\sqrt{(\varepsilon^o)^2 + (\varepsilon^f)^2} \quad , \tag{7.1}$$

where ε^o and ε^f are observation and forecast errors in observation space, respectively. Note that too many data will be rejected if the specified observation errors or spread are too small, and vice versa.

In terms of radar data thinning (cf. Section 6.8), for each 5 km in radial distance and for each 5 degree in azimuth one bin has been selected. An adaptive covariance inflation is utilized with an initial value of 1.05 and ranges from 0.5 to 3.0 (cf. Subsection 6.6.2).

7.2. Assimilation experiments

Tab. 7.1 lists four experiments performed to be discussed and evaluated in this section. At first, 4D-LETKF is first tested in E_0 by assimilating the conventional observations from AIREP, DRIBU, PILOT and SYNOP (see Chapter 1). During the chosen time, there are no TEMP data available. In terms of localization, a uniform localization strategy is applied to all types of observations. That is, horizontal localization length is chosen to 100 km, and for the vertical localization different lengths are set to different heights due to the variation of observation density with height: it begins with 0.075 log hPa at the surface level and linearly increases to 0.5 log hPa on the top level. Next, E_1 shares the same configurations as E_0 but assimilates Doppler velocity measured from the radar network of DWD as additional data by using the radar forward operator

[1] In the future we want to initialize the ensemble members by multiplying a gauss-distributed random vector with an error covariance matrix of 3D-VAR in GME-LETKF and then interpolating the resulting boundary data onto the COSMO-model.

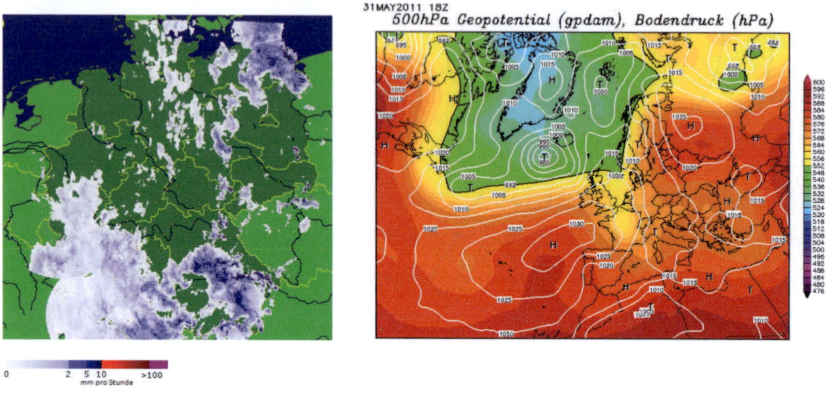

Fig. 7.1.: Left column:: precipation rate in mmh^{-1} (see color bar) on 31 May, 2011, derived from the radar network of DWD and made available form www.Niederschlagradar.de; **Right column:** geopotential in gpdam (see color bar) and surface pressure at 18:00 UTC, cited from www.Wetterzentrale.de

with simplest configurations $\mathscr{H}_{radar}$. Considering much higher resolutions of radar data than the other data, a stronger localization is applied to radar data in E_2, with the horizontal localization length cut down to 20 km. As mentioned in Chapter 5, we are also very interested in the influence and efficiency of $\mathscr{H}_{radar}$ with full configurations (compared to E_2) in the framework of data assimilation. For this reason, configurations of $\mathscr{H}_{radar}$ are upgraded to the full degree in E_3 (cf. E_4 in Chapter 5). The results of experiments are presented in the following.

	Observation		Horizontal localization [km]		
Exp.	Conventional	v_r	Conventional	v_r	Configurations of $\mathscr{H}_{radar}$
E_0	active	passive	100	100	simplest
E_1	active	active	100	100	simplest
E_2	active	active	100	20	simplest
E_3	active	active	100	20	full

Tab. 7.1.: Description of data assimilation experiments. passive: read but not assimilated, active: read and assimilated; $\mathscr{H}_{radar}$ represents the radar forward operator. Simplest and full respectively refer to the configurations of $\mathscr{H}_{radar}$ in E_0 and E_4 in Chapter 5.

Fig. 7.2 shows observation errors, RMSE, spread of prior and posterior ensembles at 21:00 UTC in E_0. These statistics are evaluated on the mandatory levels from 1000 to 100 hPa at intervals of 100 hPa with respect to the observations AIREP (the first three subfigures) and from 0 to 20000 m at intervals of 2000 m with respect to Doppler observations (the lowermost subfigure). For u, v and T, $RMSE^a$ is smaller than $RMSE^f$ throughout the entire depth of the model, which indicates that 4D-LETKF is able to extract information from observations. But we also recognize that, on one hand, observation errors are obviously set too high, on the other hand, $Spread^f$ is considerably smaller than $RMSE^f$, both of which make the model believe it performs better than it really does and thus underestimate the observed information (remember that the observation errors and spread represent the assumed validity of observations and the model, respectively). As stated in Section 6.1, tuning the observation errors is a difficult issue and will not be handled in this study. Instead, we would like to focus on the lack of spread, terminologically called underdispersion, which is usually caused by underdisturbed initial conditions, no (sufficient) consideration of model errors and finite ensemble size (Buizza et al., 2005). For Doppler velocity, $RMSE^a$ is even larger than $RMSE^f$ up to 2000 m. Since $RMSE^a$ (in the lowermost subfigure of Fig. 7.2) represents the difference between observations of v_r and the analysis by assimilating the conventional data, the phenomenon of $RMSE^a \geq RMSE^f$ suggests a disparity between conventional and radar observations at lower levels.

Fig. 7.3 illustrates the preliminary results of assimilating Doppler velocity in E_1. One can clearly see that $Spread^f$ of u, v, T and v_r is considerably reduced. From a positive point of view, this confirms the capability of assimilating radar data, but the negative effect is overconfidence of the model on itself and ignorance of observations. One should also notice that $RMSE^a$ of u, v and T has been hardly improved or even changed for the worse against $RMSE^f$ while $RMSE^a$ of v_r becomes smaller than $RMSE^f$ (cf. v_r in Fig. 7.2). This is because that radar data are much denser than AIREP and they represent different scales of correlations (radar data for small-scale correlations and AIREP for large-scale correlations). By applying a same large-scale localization length to both data the local analysis is essentially attracted to radar data and downgrades the role of AIREP. In anticipation it should be noted that this also explains the finding in Fig. 7.5, where $RMSE^f$ and $Bias^f$ of E_1 are often larger than those of E_0 for u, v and T. As previously stated, an effective way to increase spread and to reduce amount of active data used for local analysis is application of a stricter localization length, that is what happens in E_2.

Fig. 7.4 reveals the positive influences of using separate localization lengths for conventional and radar data. $Spread^f$ has been increased and $RMSE^a$ lies considerably beneath $RMSE^f$, which means that the data assimilation system can now better tune the observations into the model. When comparing E_1 and E_2 (see Fig. 7.6), $RMSE^f$ and $Bias^f$ of E_2 are now smaller than those of E_1 in general.

What is not shown here are the statistical results of E_3 with a full upgrade of $\mathscr{H}_{radar}$. This is because of its similar results as E_2, which can be explained as follows. In order to clean the data before assimilation, there is a general data preprocessing procedure that saves only the data, for which both observed and simulated data are present, in the input files of data assimilation. On the other hand, recall that the most powerful update regarding Doppler velocity is weighting by reflectivity that reduces the coverage of Doppler velocity to the area where the simulated reflectivity exists (cf. Subsection 5.2.3, the others updates like online beam propagation (e.g., SODE) and beam smoothing show marginal effects). In fact, this effect is more or less implicitly included in the data preprocessing procedure and thus we could not see the benefits brought by upgrading $\mathscr{H}_{radar}$.

Figs. 7.7, 7.8 and 7.9 illustrate the analysis ensemble mean, $Spread^a$ of temperature T and horizontal wind $\vec{v}_h := (u \;\; v)^T$ at height of 5000 m in E_0, E_1 and E_2, respectively. In E_0, $Spread^a$ is relative large close to the boundary of the model domain and small in the inner area (see Figs. 7.7b and 7.7d) because of different densities of observations. In terms of mean, the temperature T increases gradually from the west to the east in general but in the south we can observe an Alps-shaped area (cf. Fig. 2.1b) of lower temperature (see Fig. 7.7a), which results from the terrain following vertical coordinates that are actually higher than 4000 m in the area of the Alps and thus are associated with lower temperature than the vicinity. In Fig. 7.7c, the wind $\vec{v}_h$ blows from the northwest at the west boundary and changes quickly to the south wind, due to a narrow trough present at the west boundary (see Fig. 7.1).

Compared to Fig. 7.7, the most obvious differences in Fig. 7.8 are the aforementioned reduction of spread, caused by assimilation of the highly dense radar data (see Figs. 7.8b and 7.8d). With respect to the mean, the overall picture of Fig. 7.7 is reproduced by E_1 with some differences. For instance for $\vec{v}_h$ in area of $[4W,2W] \times [2S,2N]$ of Fig. 7.8c, E_1 produces considerably higher wind speeds than E_0. For T only few differences can be seen.

In Fig. 7.9 for E_2, spread has been effectively enhanced by the special treatment of localization length for radar data (see Figs. 7.9b and 7.9d), but only small changes occur with respect to the mean.

In an operational context, we are also interested in the total wall-clock elapsed time of each experiment, as shown in Table 7.2. It mainly consists of two parts: the elapsed time spent by the COSMO-model, including the radar forward operator, and the elapsed time consumed by 4D-LETKF. As already noticed in Subsection 6.6.3, the algorithm of 4D-LETKF comprises several steps, but only those steps that are responsible for the time differences are accounted for here, which are Setup and Grid loop. The former one arranges the observation vector $\mathbf{y}^o$, specifies the observation error covariance matrix $\mathbf{R}$ and conducts the quality control (i.e., Eq. (7.1)). The latter one computes the analysis covariance matrix $\mathbf{P}^a$ on each grid point of a coarse model grid[2] and save them.

In Table 7.2, we can clearly see an acute increase (601.47%) from E_0 to E_1. The reason for that is quite plausible. Due to the consideration of radar data, more efforts have to be made in Setup to treat $\mathbf{y}^o$, $\mathbf{R}$ and quality control. Furthermore, the local analyses have to consider far more observations when estimating $\mathbf{P}^a$ and thus consume more computational time in 4D-LETKF. In E_2, the time increase reduces significantly to 281.87%, which is attributable to the stronger localization that makes the local

[2]For reasons of time, the analysis covariance matrix are first computed on a coarse model grid and then interpolated onto the fine grid.

analyses have much less observations to deal with and save the computational time in Grid loop. In E_3, we first see a moderate time increase to 640.63 s in COSMO due to the upgrade of configurations of the radar forward operator (cf. Section 5), however, as mentioned before, the use of weighting by reflectivity trims the number of observations, which accelerates the process of 4D-LETKF and compensates the increased time in COSMO. Combining all these effects, E_3 requires just 110.27% more time than E_0 but much less time than E_2.

		4D-LETKF [s]			Total [s]	
Exp.	COSMO [s]	Setup	Grid loop	Total	COSMO + 4D-LETKF	Increase [%]
E_0	453.06	19.93	114.04	1214.3	1667.36	Reference
E_1	466.88	1251.27	8899.41	11229.13	11696.01	601.47%
E_2	438.25	1139.14	3655.67	5928.95	6367.20	281.87%
E_3	640.63	928.49	914.36	2865.32	3505.95	110.27%

Tab. 7.2.: Wall-clock elapsed time distribution of data assimilation experiments. **Fifth column**: total elapsed time of 4D-LETKF; **Sixth column**: total elapsed time of the experiment (i.e., COSMO + 4D-LETKF); **Seventh column**: Time increase of each experiment in reference to E_0.

7.3. Summary

In this chapter we have reviewed the performance of 4D-LETKF with a convective case through short assimilation cycles. Note that these analyses are preliminarily since critical parts of the assimilation scheme are still under construction. Therefore, the focus of experiments has been laid on the technical aspects of 4D-LETKF. The capability of 4D-LETKF to assimilate conventional and radar data was shown in E_0 and E_1, respectively. But there are two critical issues arising in E_1. First, the spread dropped dramatically, which made the model underweight the observed information. Second, RMSE and bias of u, v and T with respect to AIREP became even worse. These issues were attributable to the assimilation of highly dense radar data and relative weak localization strength. These issues were then defused in E_2 by using a shorter localization length, where the spread was successfully increased and RMSE and bias

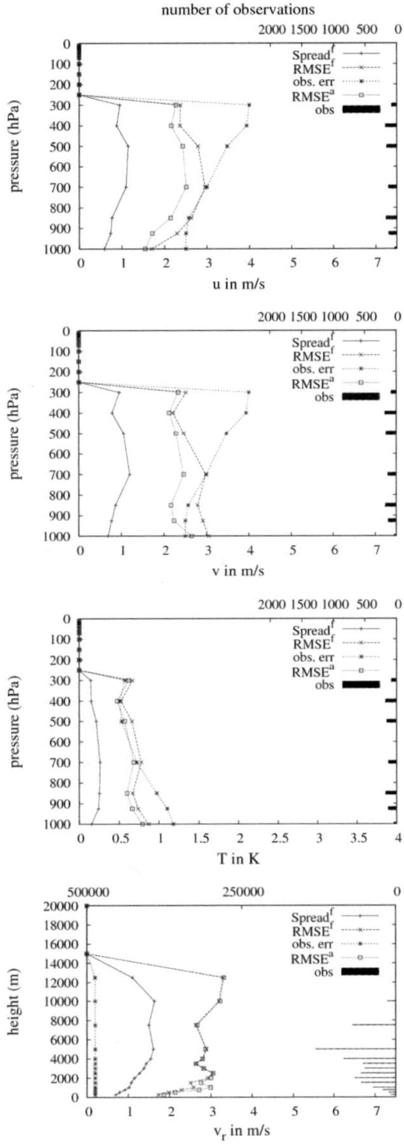

Fig. 7.2.: Various error statistical parameters $Spread^f$, $RMSE^f$, obs. err and $RMSE^a$ for 21:00 UTC in E_0 as indicated by different notations, concerning the horizontal velocity components u and v in m/s (the two uppermost subfigures), temperature T in K (the third subfigure) as function of pressure in hPa and Doppler velocity v_r in m/s (the lowermost subfigure) as function of height in m (mandatory levels only). The right ordinate refers to the number of observations, indicated by "obs" bar (note the different scale relative to the foregoing subfigures).

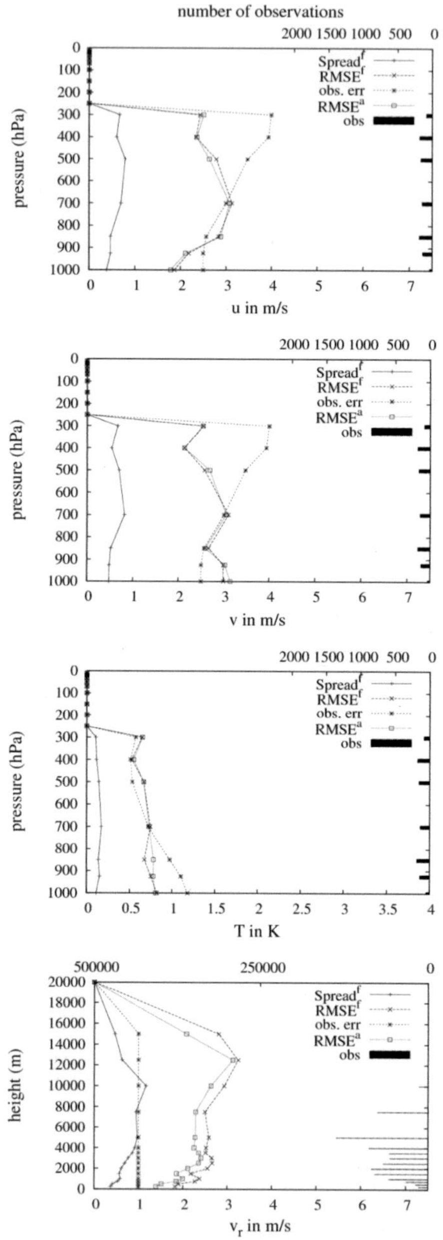

Fig. 7.3.: As Fig. 7.2 but for E_1

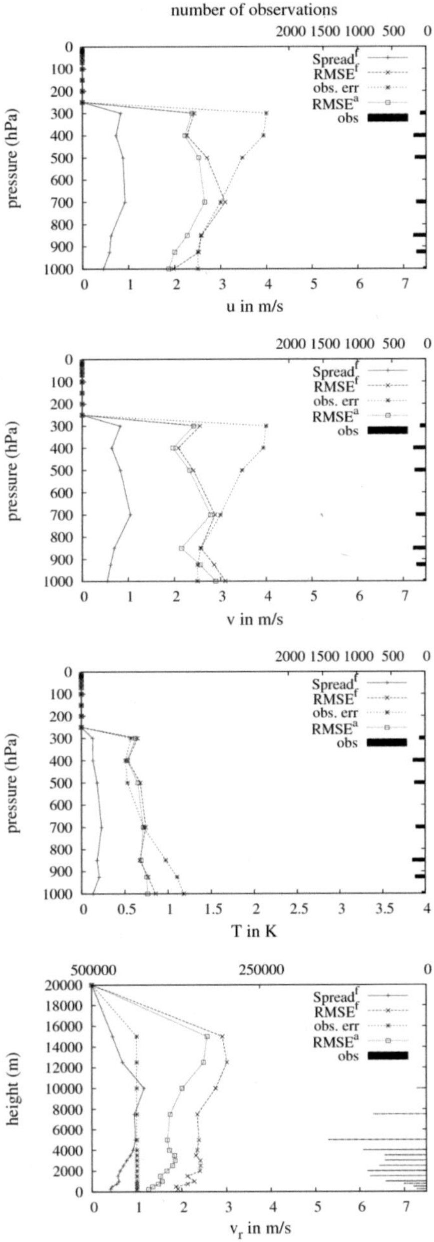

Fig. 7.4.: As Fig. 7.2 but for E_2

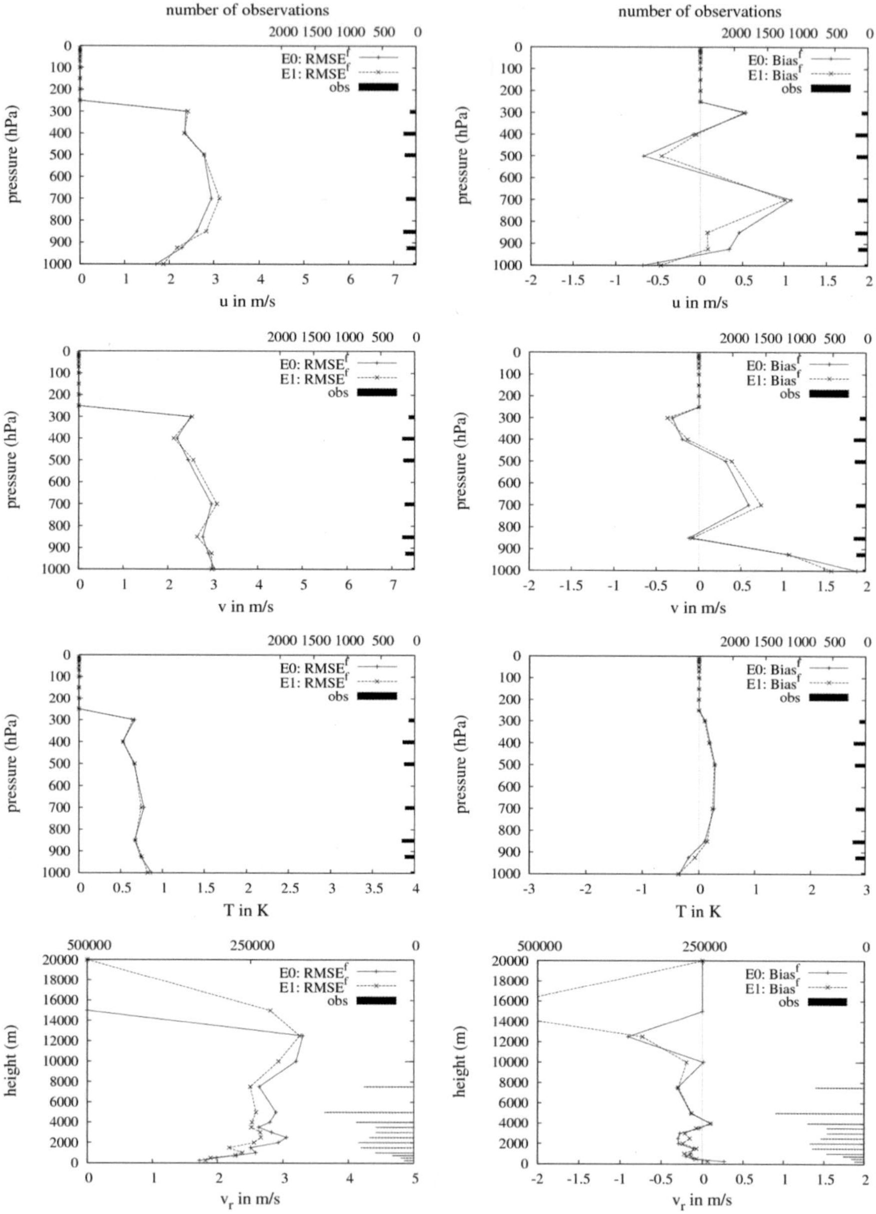

Fig. 7.5.: Comparison of $RMSE^f$ and $Bias^f$ between E_0 and E_1

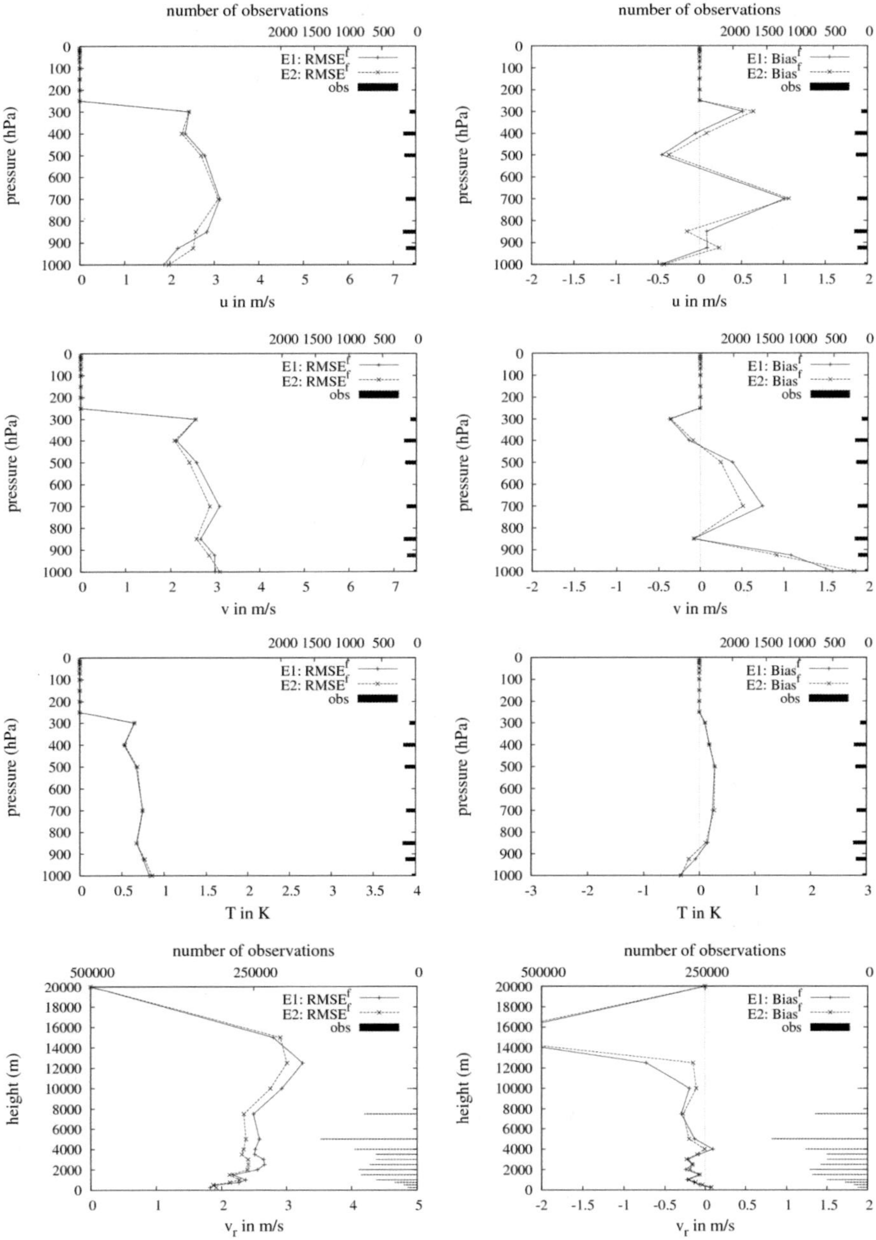

Fig. 7.6.: As Fig. 7.5 but for E_1 and E_2

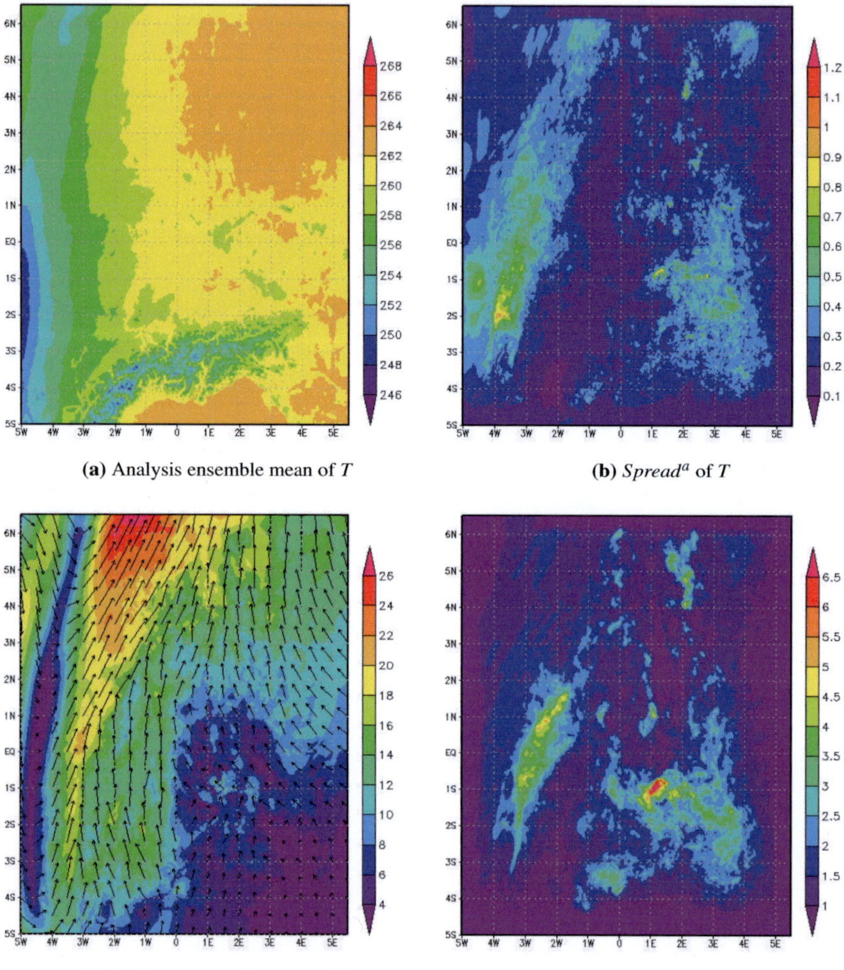

(a) Analysis ensemble mean of T

(b) $Spread^a$ of T

(c) Analysis ensemble mean of $\vec{v}_h$

(d) $Spread^a$ of $\vec{v}_h$

Fig. 7.7.: Analysis ensemble mean and $Spread^a$ at 5000 m in COSMO-DE domain (Fig. 2.1b) for 21:00 UTC in E_0. **First row:** mean and $Spread^a$ of temperature T in K; **Second row:** mean and $Spread^a$ of horizontal wind $\vec{v}_h$ in m/s

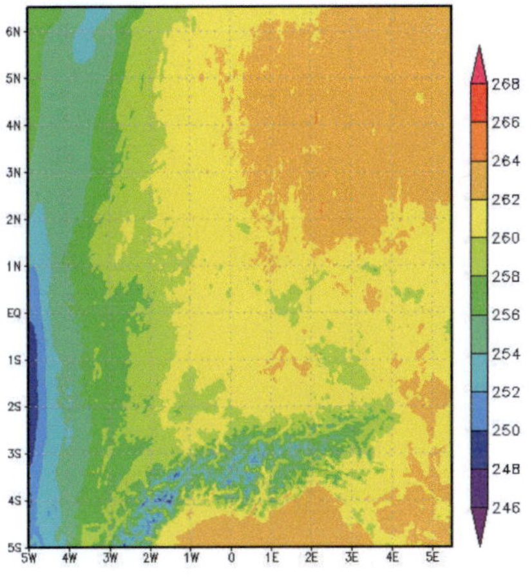

(a) Analysis ensemble mean of T

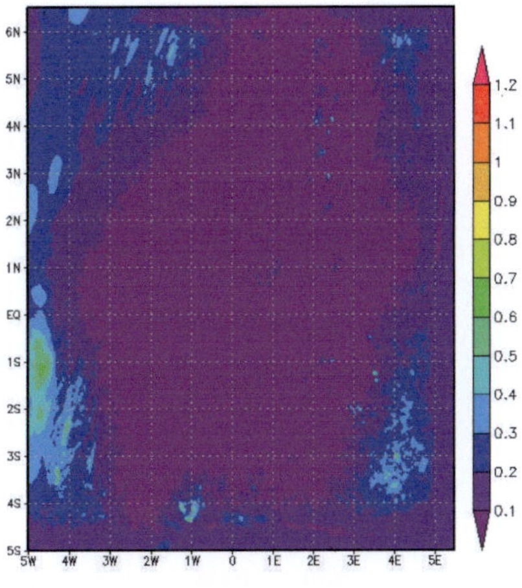

(b) $Spread^a$ of T

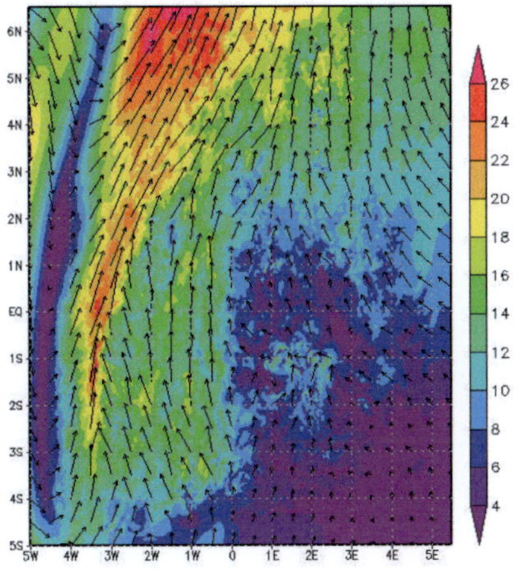

(c) Analysis ensemble mean of $\vec{v}_h$

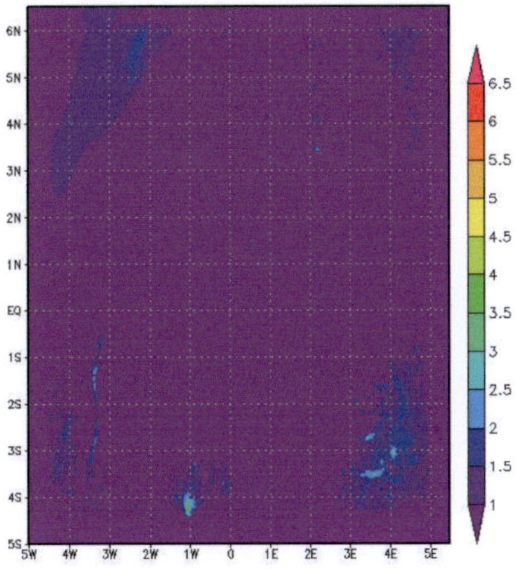

(d) $Spread^a$ of $\vec{v}_h$

Fig. 7.8.: As Fig. 7.7 but for E_1

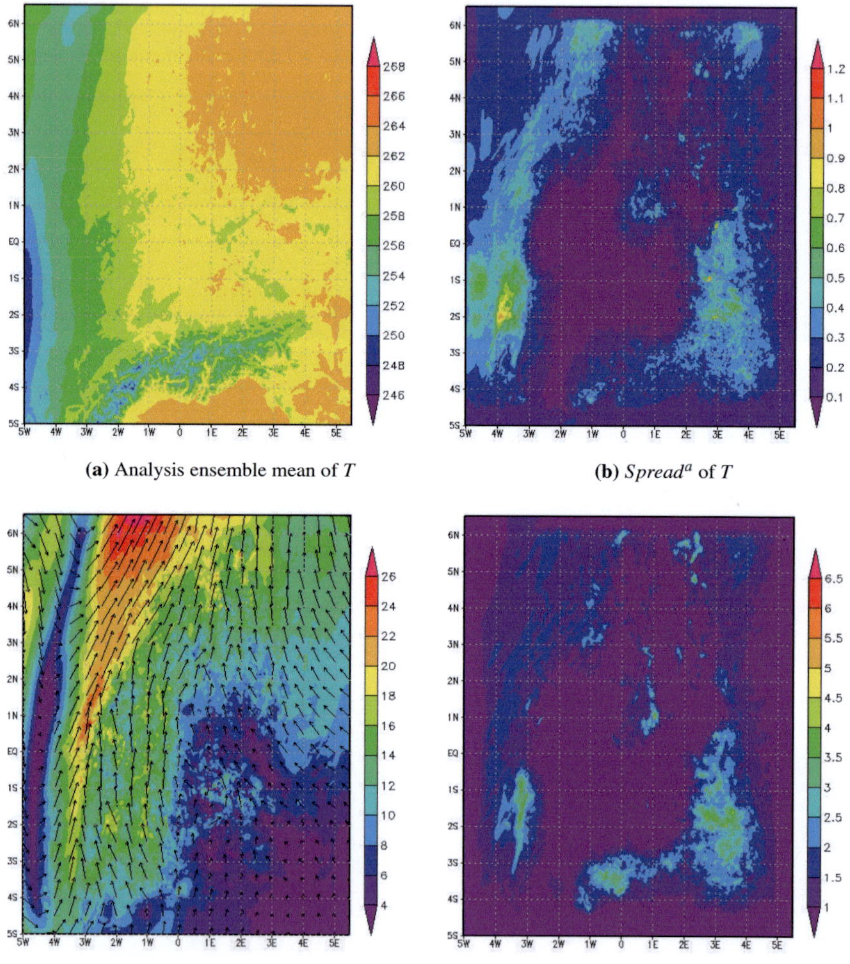

(a) Analysis ensemble mean of T

(b) $Spread^a$ of T

(c) Analysis ensemble mean of $\vec{v}_h$

(d) $Spread^a$ of $\vec{v}_h$

Fig. 7.9.: As Fig. 7.7 but for E_2

became plainly smaller. Although only few improvements in statistics have been seen with the E_3-setup, the computational time was (surprisingly) reduced against E_2, which allows the full upgrade of the radar forward operator for the operational use.

It must be recognized that these results are based on the statistics from merely one assimilation cycle. In order to obtain more convincing results, these values have to be averaged over weeks or months of cycled assimilations. Furthermore, we should choose cases that have more detailed observations, particularly TEMP, available for verification. Of cause, an effective quality control to remove gross observation errors and a realistic appraisal of the magnitude and structure of the error correlations in "good" data are also desirable. Concerning the high temporal resolution of radar observations we can consider performing the analysis step more frequently, e.g., every 15 min. Although the localization length has been cut to maintain a realistic amount of spread and to reduce RMSE, to preserve adequate continuity of analysis on adjacent points, we can not arbitrarily limit the localization length, so an optimal value of localization has to be tuned. From this point of view, an adaptive localization, depending on the density of observations, is desirable. Due to the limited ensemble size, the forecast covariance can not correctly represent large- and small-scales correlations at the same time. Motivated by this, a two-step analysis is currently under development at DWD: in the first step the conventional data are assimilated, and its results serve as first guess for the second step, performing assimilation of the additional radar data. More perspectives will be discussed in Chapter 8.

8. Conclusions and outlook

The goal of this thesis is to develop an efficient radar forward operator to be implemented into operational weather forecast COSMO[1]-model. Based on current output data of the COSMO-model, Doppler weather radar observations are calculated, which allows for a direct comparison in terms of observed radar data. With the help of the operator, the COSMO-model is able to assimilate besides usual meteorological observations the radar data. As radars provide measurements in very high spatial and temporal resolutions and DWD operates a well-distributed radar network covering the whole COSMO-DE domain, it is aimed at that the quality of operational weather forecasts, especially of short term QPFs of convective events, can be improved by means of assimilating radar data into the COSMO-model. The development work is shared with an accompanying PhD project of Dorit Jerger (Jerger et al., 2012), who concentrated on aspects of radar reflectivity, attenuation and the application of the operator as a verification tool for the cloud microphysics in the model, whereas the present thesis focused on aspects of radar beam propagation as, e.g., beam bending and broadening, programming issues as, e.g., parallelization and vectorization strategies, as well as application for data assimilation.

For the sake of flexible use, a modular operator has been designed. Each module represents a specific physical process or quantity of radar measurement, e.g., beam bending, beam broadening, Doppler velocity and reflectivity, and provides various options associated with different levels of sophistication. A series of sensitivity experiments have been conducted to find the best balance between efficiency and accuracy of modules. For beam bending, three methods have been investigated. 43ERM is a commonly applied in radar meteorology, considering climatological standard conditions; TORE is derived and modified from an existing method, characterized by total reflection and exploits Snell's law for spherically stratified media; SODE is based on a new formulation of the 2nd order ODE describing the ray propagation in spherically

[1]For the abbreviations please refer to Appendix D

stratified media based on Fermat's principle. It turned out that 43ERM works well under standard conditions or for high elevations but suffers from height overestimation in the case of superrefraction or ducting, especially for low elevations. Both TORE and SODE methods consider the actual refractivity in the atmosphere and are robust under non-standard conditions. They compute beam heights radially outwards in step of range bins instead of surface distance, which eases the implementation of the operator. But it was found that an ad hoc approach has to be imposed in TORE to determine the correct sign of local elevations at reflection points that occur under ducting conditions or negative antenna elevation. In contrast, SODE does not require such an additional constraint and shows the best stability and performance in all presented tests. Although the accuracy of online methods is also dependent on the density of aerological observations and model (vertical) resolution, the advantage of SODE will be increasingly obvious due to more dense observation networks and higher model resolution in the future. With respect to beam broadening, it is pointed out that the pulse-volume averaging generally show stronger effects in vertical than in horizontal directions and is especially necessary when the beams encounter obstacles. The averaging is numerically done by the Gauss-Legendre quadrature with variable integration points nh and nv in horizontal and vertical directions. It was recognized that for our experiments with the COSMO-model at a usual resolution, the tuple $(nh, nv) = (5, 9)$ is a reasonable choice to guarantee good averaging results.

In the stratiform and convective case studies, the ability of the model to represent different dynamical regimes has been evaluated. In terms of the stratiform event, a good agreement between model and radar observations was achieved, including the intensity and position of the event and bright band. Regarding the convective event, the model is able to catch the general evolution of the system but with some time delay. This might have contributed to disagreements in intensity and position between simulations and observations. With respect to configurations of the operator, SODE generates radar beam propagation comparable to 43ERM because no ducting conditions were prevailing in those case studies and antenna elevations are relative high. The effects of pulse-volume averaging become noticeable when beams arrive at long distances or intercept the melting level. The Mie-scattering scheme stimulates higher reflectivities than the Rayleigh approximation for partially melted particles due to the special treatment of the refractive index by Effective Medium Approximations

(EMA). Taking attenuation into account has less significance in the stratiform event but more significant impacts in the convection case. Weighting Doppler velocity by reflectivity allows for adjustment of areal coverage of Doppler velocity to the observed one. From the aspect of computational time, the most significant increase occurred when pulse-volume averaging was upgraded to $(nh, nv) = (5, 9)$. On the whole, the operator is very efficient due to its sophisticated parallelization and vectorization.

There are, however, some remaining development steps which have to be left for future work. A short-term plan includes 1) the implementation of the hydrometeor fall speed and corresponding sensitivity experiments, where improvements in high-elevation radar data are expected; 2) implementation of beam smoothing in range (up to now only in azimuthal and elevational directions), which may be important for higher-resolution models, e.g., LES (Large Eddy Simulation) models; 3) using quality control products to clean contaminated data in observations; 4) a new parallelization strategy, considering the prospective new supercomputer of DWD with thousands of processors; 5) solving the data overlapping problem of multiple radars; 6) using the operator as a validation tool for the COSMO-model (part of work has been done by my Co-PhD mate Dorit Jerger). In the long term, one ambitious goal could be to include polarimetric parameters into the operator in such a way that it is also suitable for the operational use. Several studies have shown that polarimetric radars provide more accurate information on cloud/precipitation microphysics than non-polarimetric weather radars (Brandes et al., 2002; Li and Mecikalski, 2012; Pfeifer et al., 2008), so the integration of polarimetric parameters may help to reduce the uncertainties in radar estimate of precipitation and improve radar data assimilation.

In terms of data assimilation, the preliminary results of experiments have shown that 4D-LETKF of DWD is technically able to assimilate the conventional and Doppler observations. Improvements in spread and RMSE as well as in the computational time can be attained by tuning the localization length. Moreover, it has been shown that the most accurate and expensive configurations of the radar forward operator does not encumber (actually even reduce) the computational time, which makes it feasible for the operational data assimilation. As the 4D-LETKF system is newly established, there is still much work to do. The first step may be to develop a verification tool that can quantify the added value of radar data assimilation for QPF. Alternatively, we can also assimilate observations from one radar and to use the other (radar) observations for verification. For a further insight into 4D-LETKF, a framework for OSSEs is an open working task, which allows us to apply 4D-LETKF to idealized weather systems and to better tune 4D-LETKF settings (e.g., observation errors, error correlations, localization, covariance inflation). A more sophisticated data thinning technique that avoids correlations between adjacent observations and meanwhile produces relative homogeneously distributed data is desirable. The other thinning techniques like superobing (Ramachandran et al., 2005) that combines high density clusters of data into simple but more highly weighted datum is also considerable. Since LETKF makes Gaussian assumptions over observation and forecast errors, which however are expected often non-Gaussian in reality, OSSES can also be used to quantify non-Gaussianity. Preferably within the framework of OSSEs, we would like to begin with the first experiments of assimilating reflectivity, which is expected to have positive influences not just on the analyses of the hydrometeor variables (e.g., rain, snow, hail and graupel) of the radar forward operator but also on the analyses of unobserved variables like temperature, cloud and vertical velocity. Meanwhile, the impacts on the performance of data assimilation, created by different configurations of the radar forward operator, should come under closer scrutiny. We expect to see more updating effects for reflectivity and then we can determine the optimal choice of the operator configurations in terms of data assimilation. Thereafter, more real case studies, particularly on convective events, will be performed and the potential of the assimilation for mid to short range forecast of precipitation events will be appraised.

A. DWD radar network

Since 2010, DWD has started the exchange of its Doppler C-Band weather radar network with dual-polarimetric EEC DWSR-5001C/SDP-CE radars. Currently, the network is being upgraded to new C-Band dual-polarimetric Doppler radars, and some of the radar sites will be about to change. The details of radars and their distribution are given in Tab. 1.2, where "Old" denotes the 16 radar sites before the network upgrade and "New" the 17 stations after renewal.

Name of radar station	Abbreviation	WMO NR.	Coordinates	Altitude of antenna [m]	Old	New
Berlin	BLN	10384	52.48N 13.39E	80.3	×	
Boostedt	BOO	10132	54.00N 10.05E	124.1		×
Dresden	DRS	10488	51.12N 13.77E	262.4	×	×
Eisberg	EIS	10780	49.54N 12.40E	799	×	×
Emden	EMD	10204	53.34N 7.02E	58	×	×
Essen	ESS	10410	51.41N 6.97E	185.1	×	×
Feldberg	FBG	10908	47.87N 8.00E	1517	×	×
Korbach-Rhena Flechtdorf	FLD	10440	51.31N 8.80E	623	×	×
Frankfurt-Walldo	FRI	10630	50.02N 8.56E	144.5	×	
Hamburg	HAM	10147	53.62N 10.00E	45.8	×	
Hannover	HAN	10338	52.46N 9.69E	80.75	×	×
Neuhaus	NEU	10557	50.50N 11.14E	878.5	×	×
Neuheilenbach	NHB	10605	50.11N 6.55E	585.15	×	×
			49.98N			

Offenthal	OFT	10629	8.44E	245.5		×
Prötzel	PRO	10392	52.65N 13.86E	189		×
Memmingen	MEM	10950	48.04N 10.22E	720		×
München- Fürholzen	MUC	10871	48.33N 11.61E	511.4	×	
Rostock	ROS	10169	54.18N 12.06E	36.2	×	×
Schnaupping	SNA	10873	48.17N 12.10E	724.399		×
Türkheim	TUR	10832	48.56N 9.78E	764.75	×	×
Ummendorf	UMD	10356	52.16N 11.18	183	×	×

Tab. A.1.: Information about DWD radar network

B. Numerical methods

B.1. Gauss-Legendre quadrature

In a general Gaussian quadrature rule, a definite integral $f(x)$ is first approximated over the interval $[-1,1]$ by a polynomial approximation function $g(x)$ and a known weighting function $W(x)$:

$$\int_{-1}^{1} f(x)dx = \int_{-1}^{1} W(x)g(x)dx \quad .$$

Those are then approximated by a sum of function values at specified points x_i, also called nodes, multiplied by some weights w_i:

$$\int_{-1}^{1} W(x)g(x)dx \approx \sum_{i=1}^{n} w_i g(x_i) \quad .$$

In case of Gauss-Legendre quadrature, the weighting function is $W(x) = 1$, so we can approximate the integral of $f(x)$ with:

$$\int_{-1}^{1} f(x)dx = \sum_{i=1}^{n} w_i f(x_i) \quad .$$

For this, we have to first calculate the nodes and the weights and then use them for numerical integral evaluation, which greatly speeds up the calculation compared to more simple numerical integration methods.

The n evaluation points x_i for a n-point rule are roots of nth order Legendre Polynomial $P_n(x)$. Legendre polynomials are defined by the following recursive rule:

$$\begin{aligned} P_0(x) &= 1 \quad , \\ P_1(x) &= x \quad , \\ nP_2(x) &= (2n-1)xP_{n-1}(x) - (n-1)P_{n-2}(x) \quad . \end{aligned}$$

There are also recursive equations for their derivatives:

$$P_n'(x) = \frac{n}{x^2-1}(xP_n(x) - P_{n-1}(x)) \quad .$$

The roots of those polynomials are generally not analytically solvable, so they have to be approximated numerically, for example by Newton iteration:

$$x_{n+1} = x_n - \frac{f(x_n)}{f'(x_n)} \quad .$$

The first guess x_0 for the ith root of a n-order polynomials P_n can be given by:

$$x_0 = \cos\left(\pi \frac{i - \frac{1}{4}}{n + \frac{1}{2}}\right) \quad .$$

After we get the nodes x_i, we calculate the appropriate weights by:

$$w_i = \frac{2}{(1-x_i^2)[P_n'(x_i)]^2} \quad .$$

At last, we can approximate an integral over an arbitrary interval $[a,b]$ by:

$$\int_a^b f(x)dx \approx \frac{b-a}{2} \sum_{i=1}^{n} w_i f\left(\frac{b-a}{2}x_i + \frac{a+b}{2}\right) \quad .$$

Inductively, we can deduce the 2D integral over an arbitrary interval $[a,b] \times [c,d]$ by:

$$\int_c^d \int_a^b f(x_1,x_2)dx_1dx_2 \approx \frac{d-c}{2}\frac{b-a}{2} \sum_{j=1}^{m} \sum_{i=1}^{n} w_i u_j f\left(\frac{b-a}{2}x_i + \frac{a+b}{2}, \frac{d-c}{2}x_j + \frac{d+c}{2}\right) \quad ,$$

where the interval is divided by m nodes with weights u_j, $j = 1, \ldots, m$.

B.2. One step method: fourth order explicit Runge-Kutta method

The boundary value problem of the second-order ordinary differential equation (ODE) Eq. (3.68) with boundary values Eqs. (3.70) and (3.71) is solved numerically by the one step fourth order explicit Runge-Kutta method (abbr. RK4), due to its low computational complexity and relatively reliable stability. However, to apply this method, we have to transform Eq. (3.68) to two first-order ODEs by substitution $\frac{dh}{dr} = x$:

$$
\begin{aligned}
\frac{dh}{dr} &= x \quad , \\
\frac{dx}{dr} &= x^2 \left(\frac{1}{\mathrm{n}} \frac{d\mathrm{n}}{dh} + \frac{1}{R_E + h} \right) - \left(\frac{1}{\mathrm{n}} \frac{d\mathrm{n}}{dh} + \frac{1}{R_E + h} \right) = 0 \quad ,
\end{aligned}
$$

and write them in a vector form:

$$
\frac{dy}{dr} = f(r, y) = f(r, h(r), x(r)), \text{ where } y = \begin{pmatrix} h \\ x \end{pmatrix} \quad .
$$

Numerical integrators work with discretization, i.e., one divides the integration interval $r_0 \le r \le r_e$ into subintervals $r_0 < r_1 < \cdots < r_{nra} = r_e, l = 0, 1, \ldots, nra$. $\Delta r_l = r_{l+1} - r_l$, $l = 0, \ldots, nra - 1$ is integration step. For the ease of implementation, an equidistant integration step Δr is used here, equal to pulse volume resolution.

We define $y_l := \begin{pmatrix} h(r_l) \\ x(r_l) \end{pmatrix}$. To calculate y_{l+1}, one step explicit method means that the right-hand side of Eq. (B.1):

$$
y_{l+1} = y_l + \Delta r \Phi(r_l, y_l; \Delta r) \tag{B.1}
$$

depends only on y_l, where Φ is called the increment function. Φ of RK4 is in the form:

$$\Phi = \frac{1}{6}(k_1 + 2k_2 + 2k_3 + k_4) \quad ,$$

where

$$\begin{aligned}
k_1 &= f(r_l, y_l) \quad , \\
k_2 &= f(r_l + 0.5\Delta r, y_l + 0.5k_1\Delta r) \quad , \\
k_3 &= f(r_l + 0.5\Delta r, y_l + 0.5k_2\Delta r) \quad , \\
k_4 &= f(r_l + \Delta r, y_l + k_3\Delta r) \quad .
\end{aligned}$$

Here k_2 and k_3 represent approximations to the derivative $y'(\cdot)$ at points on the solution curve, intermediate between $(r_l, y(r_l))$ and $(r_{l+1}, y(r_{l+1}))$, and $\Phi(r_l, y_l, \Delta r)$ is a weighted average of the $k_i, i = 1, \ldots, 4$.

With boundary values

$$y_0 = \begin{pmatrix} h(r=0) \\ \frac{dh}{dr}(r=0) \end{pmatrix} = \begin{pmatrix} h_0 \\ \sin\varepsilon_0 \end{pmatrix} \quad ,$$

one can now calculate y at ranges $r_l = l\Delta r, l \in \mathbb{N}_0$, and the first component of y provides the desired heights $h_l = h(r_l)$.

C. Symbols

α_0	Azimuth of radar antenna
α^m	Moisture term
β	Covariance inflation factor
c	Light speed
$C_\mathrm{p}, C_\mathrm{v}$	Heat capacities for constant pressure or volume
D	Diameter
e	Water vapor partial pressure
ε_0	Elevation of radar antenna
ε^a	Analysis error
ε^f	Forecast error
ε^o	Observation error
η	Radar reflectivity
η	Model error
$\mathbf{E}(\cdot)$	Statistical expected value
f^2	Beam weighting function
f_e^2	Effective beam weighting function
$\vec{g}$	Apparent acceleration of gravity
γ	Radio electric size
h	Height
$\mathscr{H}$	Observation forward operator
$\mathbf{H}$	Linear observation operator
$\mathbf{I}$	Identity matrix
I^x	Source/sinks of constituent x
$\mathtt{I}$	Illumination function of radar pulse volume
$\vec{J}^x$	Diffusion flux of constituent x
K	Dielectric factor

$\mathbf{K}$	Kalman gain
λ	Radar wavelength
Λ	Attenuation coefficient
ℓ	Attenuation factor
m	Complex refractive index
M	Modified refractivity
$\mathscr{M}$	Nonlinear model
$\mathbf{M}$	Tangent linear operator (Jacobian) of $\mathscr{M}$
∇	Gradient Nabla operator
n	Refractive index
N	Refractivity
$\mathscr{N}$	Distribution function
$\vec{\Omega}$	Constant angular velocity vector of earth rotation
ϕ	Horizontal angle relative to the beam axis
ϕ_3	Horizontal beamwidth
p	Pressure
p'	Perturbation pressure
$\mathfrak{p}$	Probability density function
P_r	Received power at the radar antenna
$\mathbf{P}^f$	Forecast error covariance matrix
$\mathbf{P}^a$	Analysis error covariance matrix
q^x	Mass fraction(specific content) of constituent x
Q_a	Total absorption cross section
Q_h	Diabatic heating
Q_m	Impact of changes of humidity
Q_s	Total scattering cross section
Q_t	Attenuation cross section
$\mathbf{Q}$	Model error covariance matrix
ρ	Total density of the air mixture
ρ_x	Partial density of mixture constituent x
r	Radial distance to radar antenna
r_{max}	Unambiguous range of radar
R_{eff}	Effective earth radius

R_E	Earth radius
R_v, R_d	Gas constant for water vapor and dry air
$\mathbf{R}$	Observation error covariance matrix
s	Surface distance
σ_b	Backscattering cross section
τ	Pulse duration
$\vec{\vec{\tau}}$	Stress tensor due to friction
θ	Vertical angle relative to the beam axis
θ_3	Vertical beamwidth
t	Time
T	Temperature
u	Zonal wind
v	Meridional wind
$\vec{v}$	Wind vector
v_r	Doppler velocity
$v_{r,max}$	Nyquist velocity
w	Vertical wind
w_t	Terminal fall speed of hydrometeors
$\overline{w}_t$	Average terminal fall speed of hydrometeors
W	Range weighting function
$\mathbf{x}^a$	Analysis state
$\mathbf{x}^f$	Forecast state
$\overline{\mathbf{x}}_a$	Analysis ensemble mean
$\overline{\mathbf{x}}_f$	Forecast ensemble mean
$\mathbf{x}^t$	True state
$\mathbf{X}^a$	Analysis ensemble perturbations
$\mathbf{X}^f$	Forecast ensemble perturbations
$\overline{\mathbf{y}}^f$	Observation ensemble mean
$\mathbf{y}^o$	Set of observations
$\mathbf{Y}^f$	Forecast ensemble observation perturbations
ζ	Terrain following vertical coordinate
Z	Radar reflectivity factor
Z_e	Equivalent radar reflectivity factor

D. Abbreviations and Acronyms

3/4D-VAR	Three/Four-Dimensional VARiational data assimilation
43ERM	4/3 Earth Radius Model
4D-LETKF	Four-Dimensional Local Ensemble Transform Kalman Filter
CAPS	Center for Analysis and Prediction of Storms
CFAD	Contoured Frequency with Altitude Diagrams
COSMO	COnsortium for Small-scale MOdeling
DWD	Deutscher Wetterdienst (engl. German Weather Service)
EMA	Effective Medium Approximations
GME	Global ModEl
ICAO	International Civil Aviation Organization
KENDA	Km-Scale Ensemble-Based Data Assimilation
KF	Kalman filter
EAKF	Ensemble Adjustment Kalman filter
EKF	Extended Kalman filter
EnKF	Ensemble Kalman Filter
ETKF	Ensemble Transform Kalman Filter
LES	Large Eddy Simulation
LETKF	Local Ensemble Transform Kalman Filter
LM	Local Model
MAXCAPPI	Maximum Constant Altitude Plan Position Indicator
MPI	Message Passing Interface
MSL	Mean Sea Level
NCAR	National Center for Atmospheric Research
NetCDF	Network Common Data Form
NOAA	National Oceanic and Atmospheric Administration
NWP	Numerical Weather Prediction

ODE	Ordinary Differential Equation
OSSE	Observing System Simulation Experiments
RK	Runge-Kutta method
PPI	Plan Position Indicator
PRF	Pulse Repetition Frequency
PRT	Pulse Repetition Time
PSD	Particle Size Distribution
QPF	Quantitative Precipitation Forecast
RMSE	Root Mean Square Error
SLEVE	Smooth Level VErtical
SODE	Second-order Ordinary Differential Equation
SRI	Surface Rain Intensity
TKE	Turbulent Kinetic Energy
TORE	TOtal REflection
UCAR	University Corporation for Atmospheric Research
VAD	Velocity-Azimuth Display
WRF	Weather Research and Forecasting

E. List of Figures

F. List of Tables

G. Bibliography

Anderson, J. L., 2001: An ensemble adjustment Kalman filter for data assimilation. *Mon. Wea. Rev.*, **129**, 2884–2903.

———, 2008: Spatially and temporally varing adaptive covariance inflation for ensemble filters. *Tellus*, **61A**, 72–83.

Arnold Jr., C. and C. H. Dey, 1986: Observing systems simulation experiments: Past, present, and future. *Bull. Amer. Meteor. Soc.*, **67**, 687–695.

Austin, P. M., 1987: Relation between measured radar reflectivity and surface rainfall. *Mon. Wea. Rev.*, **115**, 1053–1070.

Battan, L. J., 1973: *Radar Observations of the Atmosphere*. University of Chicago Press, Chicago.

Bean, B. R. and E. J. Dutton, 1966: *Radio Meteorology*. 92, National Bureau of Standards Monograph, U.S. Government Printing Office, Washington D.C.

Bishop, C. H., B. J. Etherton, and S. J. Majumdar, 2001: Adaptive sampling with the ensemble transform Kalman filter. Part I: Theoretical aspects. *Mon. Wea. Rev.*, **129**, 420–436.

Blahak, U., 2004: Analyse des Extinktionseffektes bei Niederschlagsmessungen mit einem C-Band Radar anhand von Simulation und Messung. PhD dissertation, Karlsruhe Insititute of Technology.

———, 2007: RADAR_MIE_LM and RADAR_MIELIB – calculation of radar reflectivity from model output. Internal report, Inst. of Meteor. and Climate research, KIT, available on request.

———, 2008a: An approximation to the effective beam weighting function for scanning meteorological radars with axissymmetric antenna pattern. *J. Atmos. Ocean. Tech*, **25**, 1182–1196.

———, 2008b: Towards a better representation of high density ice particles in a state-of-the-art two-moment bulk microphysical scheme. Extended Abstract, International Conference on Clouds and Precipitation, 7.7.-11.7.2008, Cancun, URL http://cabernet.atmosfcu.unam.mx/ICCP-2008/abstracts/Program_on_line/Poster_07/Blahak_extended_1.pdf.

Bonavita, M., L. Torrisi, and F. Marcucci, 2010: Ensemble data assimilation with the CNMCA regional forecasting system. *Quart. J. Roy. Meteor. Soc.*, **136**, 132–145.

Bouttier, F. and P. Courtier, 2002: Data assimilation concepts and methods. Meteorological Training Course Lecture Series, ECMWF, URL http://www.ecmwf.int/newsevents/training/lecture_notes/pdf_files/ASSIM/Ass_cons.pdf.

Brandes, E. A., G. Zhang, and J. Vivekanandan, 2002: Experiment in rainfall estimation with a polarimetric radar in a subtropical environment. *J. Appl. Meteor.*, **41**, 674–685.

Buizza, R., P. Houtekamer, Z. Toth, G. Pellerin, and M. Wei, 2005: A comparison of the ECMWF, MSC, and NCEP global ensemble prediction systems. *Mon. Wea. Rev.*, **133**, 1076–1097.

Burgers, G., P. J. van Leeuwen, and G. Evensen, 1998: Analysis scheme in the ensemble Kalman filter. *Mon. Wea. Rev.*, **126**, 1719–1724.

Caumont, O., 2006: An radar simulator for high-resolution nonhydrostatic models. *J. Atmos. Oceanic Technol.*, **23**, 1049–1067.

Caumont, O. and V. Ducrocq, 2008: What should be considered when simulating Doppler velocities measured by ground-based weather radars? *J. App. Meteorology and Climatology*, **47**, 2256–2262.

Chapnik, B., G. Desroziers, F. Rabier, and O. Talagrand, 2004: Properties and first application of an error-statistics tuning method in variational assimilation. *Quart. J. Roy. Meteor. Soc.*, **130**, 2253–2275.

Chen, C.-N., J.-L. Wang, C.-M. Chu, and F.-C. Lu, 2009: Ray tracing of an abnormal radar echo using geographic information system. *Defence Science Journal*, **59**, 63–72.

Cheong, B. L., R. D. Palmer, and M. Xue, 2008: A time series weather radar simulator based on high-resolution atmospheric models. *J. Atmos. Ocean. Tech*, **25**, 230–243.

Dalezios, N. R. and N. Kouwen, 1982: On the structure of homogeneous anisotropic correlation functions for real-time radar rainfall estimation. *Hydrometeorology*, American Water Resources Association, 153–157.

Davies, H., 1976: A lateral boundary formulation for multi-level prediction models. *Quar. J. Roy, Meteorol. Soc.*, **102**, 405–418.

Dazhang, T., S. G. Geotis, R. E. Passarelli, A. L. Hansen, and C. L. Frush, 1984: Evaluation of an alternating-PRF method for extending the range of unambiguous Doppler velocity. *Preprints, 22d Conf. on Radar Meteorology*, Amer. Meteor. Soc., Zurich, Switzerland, 523–527.

Doms, G. and U. Schättler, 2002: A Description of the Nonhydrostatic Regional Model LM. Part I: Dynamics and Numerics. Tech. rep., Deutscher Wetterdienst.

Doviak, R. J. and D. S. Zrnic, 1993: *Doppler Radar and Weather Observations*, Vol. 2. Academic Press, Inc., San Diego.

Dowell, D. C., F. Zhang, L. J. Wicker, C. Snyder, and N. A. Crook, 2003: Wind and temperature retrievals in the 17 May 1981 Arcadia, Oklahoma, supercell: Ensemble Kalman filter experiments. *Mon. Wea. Rev.*, **132**, 1982–2005.

Dragosavac, M., 2008: BUFR User's Guide. *ECMWF Operations Department Technical Note.*

Ducrocq, V., J. P. Lafore, and J. L. Redelsperger, 2000: Initialization of a fine-scale model for convective-system prediction: A case study. *Quart. J. Roy. Meteor. Soc.*, **126**, 3041–3065.

Evensen, G., 2003: The Ensemble Kalman Filter: Theoretical formulation and practical implementation. *Ocean Dynamics*, **53**, 343–367.

Field, P. R., A. J. Heymsfield, and A. Bansemer, 2007: Snow Size Distribution Parameterization for Midlatitude and Tropical Ice Clouds. *J. Atmos. Sci.*, **64**, 4346–4365.

Fritsch, J. M. and R. E. Carbone, 2004: Improving quantitative precipitation forecasts in the warm season. *Bull. Amer. Meteor. Soc.*, **85**, 955–965.

Fukumori, I., 2002: A partitioned Kalman filter and smoother. *Mon. Wea. Rev.*, **130**, 1370–1383.

Gal-chen, T. and R. Somerville, 1975: On the base of a coordinate transformation for the solution for Navier-Stokes equations. *J. Comput. Phys.*, **17**, 209–228.

Gao, J., K. Brewster, and M. Xue, 2006: A comparison of the radar ray path equations and approxiamtions for use in radar data assimilation. *Advances in Atmospheric sciences*, **23**, 190–198.

Gaspari, G. and S. E. Cohn, 1999: Construction of correlation functions in two and three dimensions. *Q. J. R. Meteorol. Soc.*, **125**, 723–757.

Gauthier, P., P. Courtier, and P. Moll, 1993: Assimilation of simulated wind Lidar data with a Kalman filter. *Mon. Wea. Rev.*, **121**, 1803–1820.

Gelb, A., 1974: *Applied Optimal Estimation*, Vol. 3. MIT Press, Cambridge, Massachusetts.

Gilmore, M. S., J. M. Straka, and E. N. Rasmussen, 2004: Precipitation uncertainty due to variations in precipitation particle parameters within a simple microphysics scheme. *Mon. Wea. Rev.*, **132**, 2610–2627.

Haase, G. and S. Crewell, 2000: Simulation of radar re ectivities using a mesoscale weather forecast model. *Water Resour. Res*, **36**, 2221–2231.

Haase, G. and T. Landelius, 2004: Dealiasing of Doppler Radar Velocities Using a Torus Mapping. *Journal of Atmospheric and Oceanic Technology*, **21**, 1566–1573.

Hagedorn, R., 2010: *Operational and Research Activities at ECMWF*. Presentations of the NWP-PR Training Course, ECMWF.

Halmos, P. R., 1974: Positive approximants of operators. *Indiana Univ. Math. J.*, **21**, 951–960.

Harlim, J., 2006: Errors in the intial conditions for numerical weather prediction: A study of error growth patterns and error reduction with ensemble filtering. PhD dissertation, University of Maryland.

Hartree, D. R., J. G. L. Michel, and P. Nicolson, 1946: Practical Methods for the Solution of the Equations of Tropospheric Refraction. *Meteorological factors in radio wave propagation*, The Physical Society, London, 127–168.

Hengstebeck, T., K. Helmert, and J. Seltmann, 2010: RadarQS - a standard quality control software for radar data at DWD. *Proc. 6th European Conf. on Radar in Meteorology and Hydrology (ERAD)*, Sibiu, Romania.

Heymsfield, A. J. and W. Kajikawa, 1986: An improved approach to calculating terminal velocities of platelike crystals and graupel. *J. Atmos. Sci.*, **44**, 1088–1099.

Hoke, J. E. and A. Anthes, 1976: The initialization of numerical models by a dynamic initialization technique. *Mon. Wea. Rev.*, **104**, 1551–1556.

Houtekamer, P. L. and H. L. Mitchell, 1998: Data Assimilation Using an Ensemble Kalman Filter Technique. *Mon. Wea. Rev.*, **126**, 796–811.

Houze, R. A., 1993: *Cloud Dynamics*. Academic Press, San Diego.

Hunt, B., E. Kalnay, E. Kostelich, E. Ott, D. Patil, T. Sauer, I. Szunyogh, J. Yorke, and A. Zimin, 2004: Four-dimensional ensemble Kalman filtering. *Tellus*, **56A**, 273–277.

Hunt, B. R., E. J. Kostelich, and I. Szunyogh, 2007: Efficient data assimilation for Spatiotemporal Chaos: a Local Ensemble Transform Kalman Filter. *Physica D: Nonlinear Phenomena*, **230**, 112–126.

Jacobsen, I. and E. Heise, 1982: A new economic method for the computation of the surface temperature in numerical models. *Contr. Atmos. Phys.*, **55**, 128–141.

Järvinen, H., K. Salonen, M. Lindskog, A. Huuskonen, and S. Niemelä, 2009: Doppler radar radial winds in HIRLAM. Part I: observation modelling and validation. *Tellus*, **61A**, 278–287.

Jerger, D., Y. Zeng, and U. Blahak, 2012: Efficient radar forward operator for data assimilation and model verification (1). *7th European Conference on Radar in Meteorology and Hydrology (ERAD)*, Toulouse, France.

Joss, J. and A. Waldvogel, 1990: Precipitation measurement and hydrology. *Radar in Meteorology, D. Atlas, Ed.*, Amer. Meteor. Soc., 577–606.

Kain, J. S. and J. M. Fritsch, 1993: Convective Parameterization for Mesoscale Models: The Kain-Fritsch Scheme. In: The Representation of Cumulus Convection in Numerical Models. *Meteorological Monographs*, **46**, 165–170.

Kalman, R. E., 1960: A new approach to linear filtering and prediction problems. *Trans. ASME J. Basic Engrg.*, **82**, 35–45.

Keppenne, C. L., 2000: Data assimilation into a primitive-equation model with a parallel ensemble Kalman filter. *Mon. Wea. Rev.*, **128**, 1971–1981.

Klemp, J. B. and R. Wilhelmson, 1978: The simulation of three-dimensional convective storm dynamics. *J. Atmos. Sci.*, **35**, 1070–1096.

Krajewski, W. and V. Chandrasekar, 1993: Physically Based Radar Simulation of Radar Rainfall Data Using A Space Time Rainfall Model. *J. Appl. Meteorol*, **32**, 268–283.

Li, H., E. Kalnay, and T. Miyoshic, 2009: Simultaneous estimation of covariance inflation and observation errors within an ensemble Kalman filter. *Quart. J. Roy. Meteor. Soc.*, **135**, 523–533.

Li, X. and J. R. Mecikalski, 2012: Impact of the Dual-Polarization Doppler Radar Data on Two Convective Storms with a Warm-Rain Radar Forward Operator. *Mon. Wea. Rev.*, **140**, 2147–2167.

Lindskog, M., K. Salonen, H. Järvinen, and D. B. Michelson, 2004: Doppler radar wind data assimilation with HIRLAM 3DVAR. *Mon. Wea. Rev.*, **132**, 1081–1092.

Lorenc, A. C. and C. Hammon, 1988: Objective quality control of observations using Bayesian methods. Theory, and a practical implementation. *Quart. J. Roy. Meteor. Soc.*, **114**, 515–543.

Mahafza, B. R., 2000: *Radar System Analysis and Design Using MATLAB 2nd Edition*. Chapman and Hall/CRC.

Marshall, J. S. and W. Hitschfeld, 1953: Interpretation of the fluctuating echo form randomly distributed scatterers: Part I. *Can. J. Phys*, **31**, 962–994.

May, R. M., M. I. Biggerstaff, and M. Xue, 2007: A Doppler radar emulator with an application to the dectectability of tornadic signatures. *J. Atmos. Ocean. Tech*, **24**, 1973–1996.

Maybeck, P. S., 1979: *Stochastic models, estimation, and control*, Vol. 1. Academic Press, Inc., 111 Fifth Avenue, New York, New York 10003.

McCarty, W., 2012: Cloud coverage in the joint OSSE nature run. *Mon. Wea. Rev.*, **140**, 1863–1871.

Mie, G., 1908: Beiträge zur Optik trüber Medien, speziell kolloidaler Metallösungen. *Ann. Phys.*, **25**, 377–445.

Neuper, M., 2010: Anomalous propagation: Examination of ducting conditions and anaprop events in SW-Germany. Seminar work, Kalsruhe Insititute of Technology.

Noppel, H., U. Blahak, A. Seifert, and K. D. Beheng, 2010: Simulations of a hailstorm and the impact of CCN using an advanced two-moment cloud microphysical scheme. *Atmos. Res.*, **96**, 286–301.

Oczkowski, M., I. Szunyogh, and D. Patil, 2005: Mechanisms for the development of locally low-dimensional atmospheric dynamics. *J. Atmos. Sci.*, **62**, 1135–1156.

Oguchi, T., 1983: Electromagnetic wave propagation and scattering in rain and other hydrometeors. *IEEE*, **71**, 1029–1078.

Ott, E., B. R. Hunt, I. Szunyogh, M. Corazza, E. Kalnay, P. D. J., J. A. Yorke, A. V. Zimin, and E. J. Kostelich, 2002: Exploiting local low dimensionality of the atmospheric dynamics for efficient ensemble Kalman filtering. arXiv:physics/0203058v3.

Ott, E., B. R. Hunt, I. Szunyogh, A. V. Zimin, E. J. Kostelich, M. Corazza, E. Kalnay, P. D. J., and J. A. Yorke, 2004: A local ensemble Kalman filter for atmospheric data assimilation. *Tellus A*, **56**, 415–428.

Patil, D., B. R. Hunt, E. Kalnay, J. A. Yorke, and E. Ott, 2001: Local low dimensionality of atmospheric dynamics. *Phys. Rev. Lett.*, **86**, 5878–5881.

Pfeifer, M., G. Craig, M. Hagen, and C. Keil, 2008: A polarimetric radar forward operator for model evaluation. *J. Atmos. Oceanic Technol.*, **47**, 3202–3220.

Piper, C., 2012: An Introduction to Vectorization with the Intel Fortran Compiler. http://www.polyhedron.com/web_images//documents/An_Introduction_to_Vectorization_with_Intel_Fortran_Compiler_021712.pdf.

Press, W. H., S. A. Teukolsky, W. T. Vetterling, and B. P. Flannery, 1993: *Numerical Recipes in Fortran 77*. Cambridge University Press.

Probert-Jones, J. R., 1962: The radar equation in meteorology. *Quart. J. Roy. Met. Soc.*, **88**, 485–495.

Ramachandran, R., X. Li, S. Movva, S. Graves, S. Greco, D. Emmitt, J. Terry, and R. Atlas, 2005: ntelligent Data Thinning Algorithm for Earth System Numerical Model Research and Application. *21st International Conference on Interactive Information Processing Systems (IIPS) for Meteorology, Oceanography, and Hydrology*, San Diego, USA.

Raschendorfer, M., 2001: The new turbulence parametrization of LM. *COSMO-Newsletter*, **1**, 89–97.

Reinhardt, T. and A. Seifert, 2006: A three-category ice scheme for LMK. *COSMO News Letter*, **6**, 115–120.

Rhodin, A., 2012: Feedback File Definition. http://www.cosmo-model.org/content/model/documentation/core/cosmoFeedbackFileDefinition.pdf.

Ritter, B. and J. F. Geleyn, 1992: A comprehensive radiation scheme for numerical weater prediction models with potential applications in climate simulations. *Mon. Wea. Rev.*, **120**, 303–325.

Ruffieux, D. and A. J. Illingworth, 2008: Final Report. Tech. rep., EG-CLIMET.

Rutlege, S. A. and P. V. Hobbs, 1984: The mesoscale and microscale struture and organization of clouds and precipitation in mid-latitude clouds. Part XII: A diagnostic modeling study of precipitation development in narrow cold frontal rainbands. *J. Atmos. Sci.*, **41**, 2949–2972.

Sacher, W. and P. Bartello, 2009: Sampling Errors in Ensemble Kalman Filtering. Part II: Application to a Barotropic Model. *Mon. Wea. Rev.*, **137**, 1640–1654.

Scarchilli, G., E. Gorgucci, V. Chandrasekar, and T. A. Seliga, 1993: Rainfall estimation using polarimetric techniques at C-band frequencies. *Appl. Meteor.*, **32**, 1150–1159.

Schär, C., D. Leuenberger, O. Fuhrer, D. Lüthi, and C. Girard, 2002: A New Terrain-Following Vertical Coordinate Formulation for Atmospheric Prediction Models. *Mon. Wea. Rev.*, **130**, 2459–2480.

Schraff, C., H. Reich, A. Rhodin, and A. Schomburg, 2012: Development of a LETKF for km-scale ensemble data assimilation for the COSMO-model. *Ensemble Methods Conference*, Toulouse, France.

Seifert, A. and K. D. Beheng, 2006: A two-moment cloud microphysics parameterization for mixed-phase clouds. Part I: Model description. *Meteor. Atmos. Phys.*, **92**, 45–66.

Seifert, A., A. Khain, A. Pokrovsky, and K. D. Beheng, 2006: A comparison of spectral bin and two-moment bulk mixed-phase cloud microphysics. *Atmos. Res.*, **80**, 46–66.

Seltmann, E. E. J., 2000: Clutter versus radar winds. *Phys. Chem. Earth*, **B25**, 1173–1178.

Siebren, Y. v. d. W., 2003: Ray tracing and refraction in the modified US1976 atmosphere. *Applied Optics*, **42**, 354–366.

Smoydzin, L., 2004: Theoretische und numerische Untersuchungen zur Konvektionsparametrisierung in einem Wettervorhersagemodell. Phd dissertation, University Bonn.

Snyder, C. and F. Zhang, 2003: Assimilation of simulated Doppler radar observations with an ensemble Kalman filter. *Mon. Wea. Rev.*, **131**, 1663–1677.

Sommeria, G. and J. W. Deardorff, 1977: Subgrid-Scale Condensation in Models of Nonprecipitation Clouds. *J. Atmos. Sci.*, **34**, 344–355.

Stauffer, D. R. and N. L. Seaman, 1990: Use of four-dimensional data assimilation in a limited-area mesoscale model.Part I: Experiments with synoptic-scale data. *Mon. Wea. Rev.*, **118**, 1250–1277.

Stephan, K., S. Klink, and C. Schraff, 2008: Assimilation of radar-derived rain rates into convective-scale model COSMO-DE at DWD. *Quart. J. Roy. Meteor. Soc.*, **134**, 1315–1326.

Steppeler, J., G. Doms, U. Schättler, H.-W. Bitzer, A. Gassmann, U. Damrath, and G. Gregoric, 2003: Meso-gamma scale forecasts using the non-hydrostatic model LM. *Meteor. Atmos. Phys.*, **82**, 75–96.

Stoer, J., 1999: *Numerische Mathematik I*. Springer, Berlin.

Sun, J. and N. A. Crook, 1997: Dynamical and microphysical retrieval from Doppler radar observations using a cloud model and its adjoint. Part I: Model development and simulated data experiments. *J. Atmos. Sci.*, **54**, 1642–1661.

———, 2001: Real-time low-level wind and temperature analysis using single WSR-88D data. *Wea. Forecasting*, **16**, 117–132.

Thomas, S., C. Girard, G. Doms, and U. Schättler, 2000: Semi-implicit scheme for the DWD Lokal-Modell Meteor. *Atmos. Phys.*, **75**, 105–125.

Tiedtke, M., 1989: A comprehensive mass flux scheme for cumulus parameterization in large-scale models. *Mon. Wea. Rev.*, **117**, 1779–1799.

Tippett, M. K., J. L. Anderson, C. H. Bishop, and T. M. Hamill, 2003: Ensemble Square Root Filters. *Mon. Wea. Rev.*, **131**, 1485–1490.

Tong, M. and M. Xue, 2005: Ensemble Kalman filter assimilation of Doppler radar data with a compressible nonhydrostatic model: OSS experiments. *Mon. Wea. Rev.*, **133**, 1789–1807.

Tong, M., M. Xue, and K. K. Droegemeier, 2008: Simultaneous estimation of microphysical parameters and atmospheric state with simulated radar data and ensemble square root Kalman filter. Part I: Sensitivity analysis and parameter identifiability. *Mon. Wea. Rev.*, **136**, 1630–1648.

Torn, R. D. and G. J. Hakim, 2008: Performance characteristics of a pseudo-operational ensemble Kalman filter. *Mon. Wea. Rev.*, **136**, 3947–3963.

Trier, E., S. B. Chen and K. W. Manning, 2004: A study of convection initiation in a mesoscale model using high-resolution land surface initial conditions. *Mon. Wea. Rev.*, **132**, 2954–2976.

Turton, J. D., D. A. Bennetts, and S. F. G. Farmer, 1988: An introduction to radio ducting. *Meteorol. Mag.*, **117**, 245–254.

Vetter, S., Y. Aoyama, and N. J., 1999: *RS/6000 SP: Practical MPI Programming*. IBM Redbooks.

Wang, X., C. H. Bishop, and S. J. Julier, 2004: Which is better, an ensemble of positive-negative pairs or a centered spherical simplex ensemble? *Mon. Wea. Rev.*, **132**, 1590–1605.

Weisman, M. L. and J. B. Klemp, 1982: The dependence of numerically simulated convective storms on vertical wind shear and buoyancy. *Mon. Wea. Rev.*, **110**, 504–520.

Whitaker, J. S. and T. M. Hamill, 2002: Ensemble data assimilation without perturbed observations. *Mon. Wea. Rev.*, **130**, 1913–1924.

Wicker, L. and W. Skamarock, 2002: Time-splitting methods for elastic models using forward time schemes. *Mon. Wea. Rev.*, **130**, 2088–2097.

Wilson, J. W. and E. A. Brandes, 1979: Radar measurement of rainfall-A Summary. *Amer. Meteor. Soc.*, **60**, 1048–1058.

Wu, B., M. Tong, and K. K. Droegemeier, 2000: Dynamical and microphysical retrievals from Doppler radar observations of a deep convective cloud. *J. Atmos. Sci.*, **57**, 262–283.

Xiao, Q., E. Lim, D.-J. Won, J. Sun, W.-C. Lee, M.-S. Lee, W.-J. Lee, J. Cho, Y.-H. Kuo, D. Barker, D.-K. Lee, and H.-S. Lee, 2008: Doppler radar data assimilation in KMA's operational forecasting. *Bull. Amer. Meteor. Soc.*, **89**, 39–43.

Xiao, Q. and J. Sun, 2007: Multiple radar data assimilation and short-range quantitative precipitation forecasting of a squall line observed during IHOP_2002. *Mon. Wea. Rev.*, **135**, 3381–3404.

Yuter, S. E. and R. A. Houze, 1995: Three-dimensional kinemetic and microphysical evolution of Florida cumulonimbus. Part II: Frequency distributions of vertical velocity, reflectivity, and differential reflectivity. *Mon. Wea. Rev.*, **123**, 1941–1963.

Zawadzki, I., 1973: Statistical properties of precipitation patterns. *J. Appl. Meteor.*, **12**, 459–472.

Zhang, F., Y. Weng, J. A. Sippel, Z. Meng, and C. H. Bishop, 2009: Cloud-resolving hurricane initialization and prediction through assimilation of Doppler radar observations with an ensemble Kalman filter. *Mon. Wea. Rev.*, **137**, 2105–22 125.

Acknowledgment

It would not have been possible to finish this thesis without the help and support of many nice people around me, to only some of whom I can give particular mention here.

First and foremost I offer my sincerest gratitude to my supervisor, Dr. Ulrich Blahak, who trusted me in this challenging project and introduced me to the exciting world of the radar meteorology. He has been always there for me throughout my PhD study with his patience, encouragement and brilliant ideas whilst allowing me the room to work in my own way. His wisdom, knowledge, passion in research, commitment to the highest scientific standards and hard work motivated me all the time. His support and friendship are invaluable for me on both academic and personal levels. One simply could not wish for a better supervisor.

I am deeply grateful to Prof. Klaus Beheng for his willingness to accept me as a PhD student, many stimulating discussions and the critical review of this thesis. In addition, I was delighted to attend his classes and feel his excitement in teaching. I profit highly from his profound understanding in theoretical meteorology and also appreciate his great efforts to clarify the organization of my rigorosums exam.

I want to express my gratitude to Prof. Christoph Kottmeier for kindly accepting the appointment as Co-supervisor for this thesis.

Many thanks go to Prof. Matthias Steinhauser for taking the role as my second mentor.

I would like to thank my project buddy and office mate, Dorit Jerger, not just for her contributions in implementation of extinction, this project would have truly suffered in the absence of her input, but also for her always carrying the office key for us during the lunch break for the last three years.

I am also indebted and thankful to Dr. Jan Handwerker and Gabi Klinck for their kindness and technical supports in computer. With Jan I had also numerous productive scientific discussions although he had to repeat the same stuff many times (mostly because he talked too fast).

I appreciate Malte Neuper's contributions to the simulation of online beam propagation and his readiness to answer my meteorological questions even when he was busy with his own work.

In terms of data assimilation, I really appreciate the efforts made by Dr. Andreas Rhodin and Dr. Hendrik Reich. During the last few months, I had very intensive discussions and email exchanges with Hendrik, who has always been patient with me.

Special thanks go to Dr. Daniel Leuenberger (MeteoSwiss) for supplying the initial code of the radar forward operator.

I would like to show my gratitude to Dr. Chris Synder for inviting me to the National Center for Atmospheric Research (NCAR) in USA for 3 months. I want to thank Dr. Glen Romine and Dr. Hui Liu for helping me with DART and for several fruitful discussions. I am also very thankful to Dr. Jenny Sun for introducing me to many friends and for the fun at weekends. The talks about science, life and religion with her and the other colleagues at NCAR inspired me indeed.

I owe sincere and earnest thankfulness to Graduate School for Climate and Environment (GRACE) for allowing me attend the summer school on Isle of Arran in Scotland in 2011, the foreign stay at NCAR in 2012 and participitation in 7th European Conference on Severe Storms (ECSS) 2013 in Helsinki.

My deepest gratitude goes to German Weather Service (DWD) for funding this project and academic supports.

Last, but by no means least, I would like to thank my parents for their unequivocal support and endless love throughout, as always, for which my mere expression of thanks likewise does not suffice.

Wissenschaftliche Berichte des Instituts für Meteorologie und Klimaforschung des Karlsruher Instituts für Technologie (0179-5619)

Bisher erschienen:

Nr. 1: *Fiedler, F. / Prenosil, T.*
Das MESOKLIP-Experiment. (Mesoskaliges Klimaprogramm im Oberrheintal).
August 1980

Nr. 2: *Tangermann-Dlugi, G.*
Numerische Simulationen atmosphärischer Grenzschichtströmungen über langgestreckten mesoskaligen Hügelketten bei neutraler thermischer Schichtung.
August 1982

Nr. 3: *Witte, N.*
Ein numerisches Modell des Wärmehaushalts fließender Gewässer unter Berücksichtigung thermischer Eingriffe.
Dezember 1982

Nr. 4: *Fiedler, F. / Höschele, K. (Hrsg.)*
Prof. Dr. Max Diem zum 70. Geburtstag.
Februar 1983 (vergriffen)

Nr. 5: *Adrian, G.*
Ein Initialisierungsverfahren für numerische mesoskalige Strömungsmodelle.
Juli 1985

Nr. 6: *Dorwarth, G.*
Numerische Berechnung des Druckwiderstandes typischer Geländeformen.
Januar 1986

Nr. 7: *Vogel, B.; Adrian, G. / Fiedler, F.*
MESOKLIP-Analysen der meteorologischen Beobachtungen von mesoskaligen Phänomenen im Oberrheingraben.
November 1987

Nr. 8: *Hugelmann, C.-P.*
Differenzenverfahren zur Behandlung der Advektion.
Februar 1988

Nr. 9: *Hafner, T.*
Experimentelle Untersuchung zum Druckwiderstand der Alpen.
April 1988

Nr. 10: *Corsmeier, U.*
Analyse turbulenter Bewegungsvorgänge in der maritimen atmosphärischen Grenzschicht.
Mai 1988

Nr. 11: *Walk, O. / Wieringa, J.(eds)*
Tsumeb Studies of the Tropical Boundary-Layer Climate.
Juli 1988

Nr. 12: *Degrazia, G. A.*
Anwendung von Ähnlichkeitsverfahren auf die turbulente Diffusion in der konvektiven und stabilen Grenzschicht.
Januar 1989

Nr. 13: *Schädler, G.*
Numerische Simulationen zur Wechselwirkung zwischen Landoberflächen und atmophärischer Grenzschicht.
November 1990

Nr. 14: *Heldt, K.*
Untersuchungen zur Überströmung eines mikroskaligen Hindernisses in der Atmosphäre.
Juli 1991

Nr. 15: *Vogel, H.*
Verteilungen reaktiver Luftbeimengungen im Lee einer Stadt – Numerische Untersuchungen der relevanten Prozesse.
Juli 1991

Nr. 16: *Höschele, K.(ed.)*
Planning Applications of Urban and Building Climatology – Proceedings of the IFHP / CIB-Symposium Berlin, October 14-15, 1991.
März 1992

Nr. 17: *Frank, H. P.*
Grenzschichtstruktur in Fronten.
März 1992

Nr. 18: *Müller, A.*
Parallelisierung numerischer Verfahren zur Beschreibung von Ausbreitungs- und chemischen Umwandlungsprozessen in der atmosphärischen Grenzschicht.
Februar 1996

Nr. 19: *Lenz, C.-J.*
Energieumsetzungen an der Erdoberfläche in gegliedertem Gelände.
Juni 1996

Nr. 20: *Schwartz, A.*
Numerische Simulationen zur Massenbilanz chemisch reaktiver Substanzen im mesoskaligen Bereich.
November 1996

Nr. 21: *Beheng, K. D.*
Professor Dr. Franz Fiedler zum 60. Geburtstag.
Januar 1998

Nr. 22: *Niemann, V.*
Numerische Simulation turbulenter Scherströmungen mit einem Kaskadenmodell.
April 1998

Nr. 23: *Koßmann, M.*
Einfluß orographisch induzierter Transportprozesse auf die Struktur der atmosphärischen Grenzschicht und die Verteilung von Spurengasen.
April 1998

Nr. 24: *Baldauf, M.*
Die effektive Rauhigkeit über komplexem Gelände – Ein Störungstheoretischer Ansatz.
Juni 1998

Nr. 25: *Noppel, H.*
Untersuchung des vertikalen Wärmetransports durch die Hangwindzirkulation auf regionaler Skala.
Dezember 1999

Nr. 26: *Kuntze, K.*
Vertikaler Austausch und chemische Umwandlung von Spurenstoffen über topographisch gegliedertem Gelände.
Oktober 2001

Nr. 27: *Wilms-Grabe, W.*
Vierdimensionale Datenassimilation als Methode zur Kopplung zweier verschiedenskaliger meteorologischer Modellsysteme.
Oktober 2001

Nr. 28: *Grabe, F.*
Simulation der Wechselwirkung zwischen Atmosphäre, Vegetation und Erdoberfläche bei Verwendung unterschiedlicher Parametrisierungsansätze.
Januar 2002

Nr. 29: *Riemer, N.*
Numerische Simulationen zur Wirkung des Aerosols auf die troposphärische Chemie und die Sichtweite.
Mai 2002

Nr. 30: *Braun, F. J.*
Mesoskalige Modellierung der Bodenhydrologie.
Dezember 2002

Nr. 31: *Kunz, M.*
Simulation von Starkniederschlägen mit langer Andauer über Mittelgebirgen.
März 2003

Nr. 32: *Bäumer, D.*
Transport und chemische Umwandlung von Luftschadstoffen im Nahbereich von Autobahnen – numerische Simulationen.
Juni 2003

Nr. 33: *Barthlott, C.*
Kohärente Wirbelstrukturen in der atmosphärischen Grenzschicht.
Juni 2003

Nr. 34: *Wieser, A.*
Messung turbulenter Spurengasflüsse vom Flugzeug aus.
Januar 2005

Nr. 35: *Blahak, U.*
Analyse des Extinktionseffektes bei Niederschlagsmessungen mit einem C-Band Radar anhand von Simulation und Messung.
Februar 2005

Nr. 36: *Bertram, I.*
Bestimmung der Wasser- und Eismasse hochreichender konvektiver Wolken anhand von Radardaten, Modellergebnissen und konzeptioneller Betrachtungen.
Mai 2005

Nr. 37: *Schmoeckel, J.*
Orographischer Einfluss auf die Strömung abgeleitet aus Sturmschäden im Schwarzwald während des Orkans „Lothar".
Mai 2006

Nr. 38: *Schmitt, C.*
Interannual Variability in Antarctic Sea Ice Motion: Interannuelle Variabilität antarktischer Meereis-Drift.
Mai 2006

Nr. 39: *Hasel, M.*
Strukturmerkmale und Modelldarstellung der Konvektion über Mittelgebirgen.
Juli 2006

Ab Band 40 erscheinen die Wissenschaftlichen Berichte des Instituts für Meteorologie und Klimaforschung bei KIT Scientific Publishing (ISSN 0179-5619).
Die Bände sind unter www.ksp.kit.edu als PDF frei verfügbar oder als Druckausgabe bestellbar.

Nr. 40: *Lux, R.*
Modellsimulationen zur Strömungsverstärkung von orographischen Grundstrukturen bei Sturmsituationen. (2007)
ISBN 978-3-86644-140-8

Nr. 41: *Straub, W.*
Der Einfluss von Gebirgswellen auf die Initiierung und Entwicklung konvektiver Wolken. (2008)
ISBN 978-3-86644-226-9

Nr. 42: *Meißner, C.*
High-resolution sensitivity studies with the regional climate model COSMO-CLM. (2008)
ISBN 978-3-86644-228-3

Nr. 43: *Höpfner, M.*
Charakterisierung polarer stratosphärischer Wolken mittels hochauflösender Infrarotspektroskopie. (2008)
ISBN 978-3-86644-294-8

Nr. 44: *Rings, J.*
Monitoring the water content evolution of dikes. (2009)
ISBN 978-3-86644-321-1

Nr. 45: *Riemer, M.*
Außertropische Umwandlung tropischer Wirbelstürme: Einfluss auf das Strömungsmuster in den mittleren Breiten. (2012)
ISBN 978-3-86644-766-0

Nr. 46: *Anwender, D.*
Extratropical Transition in the Ensemble Prediction System of the ECMWF: Case Studies and Experiments. (2012)
ISBN 978-3-86644-767-7

Nr. 47: *Rinke, R.*
Parametrisierung des Auswaschens von Aerosolpartikeln durch Niederschlag. (2012)
ISBN 978-3-86644-768-4

Nr. 48: *Stanelle, T.*
Wechselwirkungen von Mineralstaubpartikeln mit thermodynamischen und dynamischen Prozessen in der Atmosphäre über Westafrika. (2012)
ISBN 978-3-86644-769-1

Nr. 49: *Peters, T.*
Ableitung einer Beziehung zwischen der Radarreflektivität, der Niederschlagsrate und weiteren aus Radardaten abgeleiteten Parametern unter Verwendung von Methoden der multivariaten Statistik. (2012)
ISBN 978-3-86644-323-5

Nr. 50: *Khodayar Pardo, S.*
High-resolution analysis of the initiation of deep convection forced by boundary-layer processes. (2012)
ISBN 978-3-86644-770-7

Nr. 51: *Träumner, K.*
Einmischprozesse am Oberrand der konvektiven atmosphärischen Grenzschicht. (2012)
ISBN 978-3-86644-771-4

Nr. 52: *Schwendike, J.*
Convection in an African Easterly Wave over West Africa and the Eastern Atlantic: A Model Case Study of Hurricane Helene (2006) and its Interaction with the Saharan Air Layer. (2012)
ISBN 978-3-86644-772-1

Nr. 53: *Lundgren, K.*
Direct Radiative Effects of Sea Salt on the Regional Scale. (2012)
ISBN 978-3-86644-773-8

Nr. 54: *Sasse, R.*
Analyse des regionalen atmosphärischen Wasserhaushalts unter Verwendung von COSMO-Simulationen und GPS-Beobachtungen. (2012)
ISBN 978-3-86644-774-5

Nr. 55: *Grenzhäuser, J.*
Entwicklung neuartiger Mess- und Auswertungsstrategien für ein scannendes Wolkenradar und deren Anwendungsbereiche. (2012)
ISBN 978-3-86644-775-2

Nr. 56: *Grams, C.*
Quantification of the downstream impact of extratropical transition for Typhoon Jangmi and other case studies. (2013)
ISBN 978-3-86644-776-9

Nr. 57: *Keller, J.*
Diagnosing the Downstream Impact of Extratropical Transition Using Multimodel Operational Ensemble Prediction Systems. (2013)
ISBN 978-3-86644-984-8

Nr. 58: *Mohr, S.*
Änderung des Gewitter- und Hagelpotentials im Klimawandel. (2013)
ISBN 978-3-86644-994-7

Nr. 59: *Puskeiler, M.*
Radarbasierte Analyse der Hagelgefährdung in Deutschland. (2013)
ISBN 978-3-7315-0028-5

Nr. 60: *Zeng, Y.*
Efficient Radar Forward Operator for Operational Data Assimilation within the COSMO-model. (2014)
ISBN 978-3-7315-0128-2